SPIRE
CCEED
PROGRESS

Complete
Mathematics
or Cambridge IGCSE®
Teacher Resource Pack

fth edition

Extended

For the
updated
syllabus

n Bettison

athew Taylor

xford excellence for Cambridge IGCSE®

OXFORD

UNIVERSITY PRESS

Great Clarendon Street, Oxford, OX2 6DP, United Kingdom

Oxford University Press is a department of the University of Oxford.
It furthers the University's objective of excellence in research,
scholarship, and education by publishing worldwide. Oxford is a
registered trade mark of Oxford University Press in the UK and in
certain other countries

British Library Cataloguing in Publication Data

Data available

ISBN: 978-0-19-842807-7

10 9 8 7 6 5 4 3 2 1

Printed in Great Britain by Ashford Colour Press Ltd. Gosport

Acknowledgements

® IGCSE is the registered trademark of Cambridge Assessment International
Education.

Cambridge Assessment International Education bears no responsibility for the
example answers to questions taken from its past question papers which are
contained in this publication.

Cover photo: Phaif/Dreamstime

INTRODUCTION

This guide is designed to provide a structure for the teaching of the **Cambridge IGCSE Mathematics** syllabus at **Extended** level. There are a number of possible approaches suggested within the lessons but it does not give a definitive programme of study.

Students following the Extended syllabus are expected to have a *significant amount of prior knowledge* for many of the topics. As a result of this, many of the simpler topics and exercises are referred to only in passing.

Each **lesson** (or series of lessons) usually covers one or two key aspects of the syllabus and refers to the relevant pages of the Extended student textbook and associated exercises. It is expected that most of the main teaching points will be covered in the body of a single lesson, but that further lessons may be needed to consolidate and refresh the learning. Hence, each lesson is a guide to the time it might take to *teach* the topic rather than a prescriptive scheme of work.

Many of the **starter activities** are designed to get the students engaging with mathematics and are often only loosely connected with the body of the lesson. Often they are activities which are designed to refresh a fundamental skill required to achieve success in the topic(s) covered (for example, calculation skills). They are suggestions and can be modified or interchanged according to the needs of the specific group of students.

The **main lesson commentary** is designed to guide you through the topic(s), suggesting a possible order of teaching and picking up on certain key points which should be noted. These notes are not prescriptive and may often suggest activities that are not appropriate to either the teaching style of the teacher or the learning styles of the students. Flexibility of approach is to be expected. Suggested consolidation activities and/or extension activities are also noted where relevant.

The **exercise commentary** guides you through the various exercises associated with the topic. It suggests the types of question students of differing abilities should be attempting. Key misconceptions are also addressed. The nature and extent of the textbook exercises allow flexibility when students are consolidating, and a differentiated approach is encouraged.

The **plenary activities** are designed to summarise learning, but again these are not prescriptive. Card-matching activities, pairs and group work may not be appropriate, and a simple question and answer session or short test can be used instead. The aim is to suggest *possible* activities that might engage the learners in a different way.

There are many references to 'mass response' tools such as *mini-whiteboards* and *response cards*. Many schools do not have access to this type of equipment and therefore alternative methods have also been suggested, such as the students writing the answers in exercise books and checking them at the end.

There are references to the use of *overhead projectors, interactive whiteboards,* and other computer equipment. Not all schools will have access to these facilities, so all of the activities which suggest them can be easily modified to use pencil or prepared worksheets. Suggestions on resources:

- Response cards such as 'True' or 'False' can easily be made in class by the students by taking a piece of stiff card (A5 size or equivalent) and putting a 'T' on one side and an 'F' on the other. They work even better if the 'T' is one colour and the 'F' a different colour.

- Card-matching activities can be prepared beforehand and laminated (if facilities exist). They can be placed into a 'bank' of resources for use by other teachers and in future years.

- Mini-whiteboards can also be made by laminating stiff card, but dry-wipe pens and erasers will still be needed. Mini-chalkboards make an excellent alternative if the school has access to them.

- Scientific calculators should be available for all students following this course but if they are not, expected approximations to π are given and trigonometrical tables should be made available at the appropriate points.

The online components are designed to supplement and support the teaching of this course. There are supplementary worksheets for use at certain points during the course, practice papers, sample worked solutions and a glossary of key terms.

Access your support website at
www.oxfordsecondary.com/9780198428077

ABOUT CAMBRIDGE IGCSE MATHEMATICS

Candidates can be entered for one of two different curriculums. The table below indicates the method of assessment for both the core and extended curriculum:

Syllabus 0580

Core curriculum (grades available C–G):	Extended curriculum (grades available A*–E):
Paper 1, Short-answer questions	**Paper 2, Short-answer questions**
1 hour, 35% of total marks	1 hour 30 minutes, 35% of total marks
Paper 3, Structured questions	**Paper 4, Structured questions**
2 hours, 65% of total marks	2 hours 30 minutes, 65% of total marks

Notes

- Candidates should have an electronic calculator for all papers. Algebraic or graphical calculators are not permitted. Three significant figures will be required in answers except where otherwise stated.

- Candidates should use the value of π from their calculators if their calculator provides this. Otherwise, they should use the value of 3.142 given on the front page of the question paper only.

- Tracing paper may be used as an additional material for each of the written papers.

CONTENTS

CHAPTER 1
NUMBER

Lesson 1 – Decimals

Textbook pages 2–3

Expected prior knowledge Students should, at this level, have a working knowledge of the rules of arithmetic when applied to whole numbers and decimals fractions. This lesson provides a good opportunity to revise these methods.

Objectives
E1.8: Use the four rules for calculations with decimals, including correct ordering of operations and use of brackets.

Starter
Give students a range of questions where they are asked to multiply and divide by powers of 10. This could be done as a question and answer session or they could write the answers in their books for checking at the end.

Lesson commentary
- Students should be familiar with the basic rules of arithmetic and the idea of decimal place value. Use this as a starting point to demonstrate the methods for decimal calculations. Use a range of examples.

- Include examples of decimal addition and subtraction as well as multiplication and division, emphasising all the while the need to check the size of the answer relative to the numbers in the question. Incorrectly positioning the decimal point in the final answer is a common problem which can be overcome by checking the size of the answer. Students should be advised to first estimate the answer by approximating the numbers before comparing

The support worksheet for Chapter 1 may be useful to support lessons 2, 7, 8 and 9.

The challenge worksheet for Chapter 1 may be useful to support lesson 4.

Exercise commentary
Exercise 1 provides routine practice in applying the rules of arithmetic to decimals. **Questions 32, 33** and **35 to 40** require the correct use of the rules of BIDMAS so ensure students take this into account (students should have a working knowledge of the rules of BIDMAS from previous work but if they do not, these questions could be left out).

Exercise 2 is a problem-solving exercise which can be done at this point for interest or set as extension work or homework. Alternatively, a further lesson can be used to enable students to work on these applied problems as a class.

this to their actual answer. Simple arithmetical mistakes cannot be identified in this way but the checking of answers on a calculator could be encouraged.

- As an extension exercise, students could be encouraged to set their own decimal arithmetic questions for a partner to solve (ensuring that they can solve them first).

- If necessary, a second lesson can be used to provide students with additional time to consolidate this work.

Plenary
Give the students a starting number and then a series of operations involving decimal numbers. Students then work out each individual step before recording the final answer. An example could be:
$4.2 \times 3 - 1.3 + 6.7 \div 1.5 \ (= 12)$.

Lesson 2 – Fractions

Textbook pages 4–7

Expected prior knowledge Students should be familiar with simple equivalences and the methods for carrying out arithmetic with fractions.

Objectives

E1.5: Use the language and notation of simple vulgar and decimal fractions and percentages in appropriate contexts. Recognise equivalence and convert between these forms.

E1.6: Order quantities by magnitude.

E1.8: Use the four rules for calculations with decimals and fractions (including mixed numbers and improper fractions), including correct ordering of operations and use of brackets.

Starter

Simple fraction, decimal and percentage conversions: Ask students to write down the decimal and percentage equivalents of a half, a quarter, three quarters, a fifth and a tenth. The difficulty could be increased to include a number of hundredths or fractions such as one third and one eighth. They could write these down on mini-whiteboards or in their books.

Lesson commentary

- Assuming students are familiar with arithmetic involving fractions, ask them to write down a description of the methods for adding, subtracting, multiplying and dividing fractions as if they were writing a list of instructions for a classmate. Ask them to provide a worked example of each.

- Students should also be familiar with simple fraction and decimal equivalents (see starter activity) but should then be reminded of the more general methods for converting between forms.

- Demonstrate each of the techniques using a range of suitable examples taken from the textbook or other source.

- Decimal to fraction conversions can be done by writing the decimal over the appropriate power of ten and then simplifying. Provide a number of examples of this and then allow students time to consolidate if necessary.

- When converting from fractions to decimals, emphasise that the numerator goes *inside* the division and the denominator *outside*. Add extra zeros as appropriate and retain the position of the decimal point. Students should then treat the division the same as any other (short) division problem.

- Allow time for the students to practise converting between the forms. A second lesson can be used to ensure all students have fully consolidated both fraction arithmetic and fraction/decimal conversions.

- Model converting from recurring decimals to fractions using the examples in the textbook or elsewhere. Emphasise the importance of multiplying by the right power of 10 to shift the digits in order to get cancelling out on subtraction.

Exercise commentary

Questions 1 to 15 in exercise 3 are basic practice questions while **questions 16 to 21** introduce mixed numbers. Ensure methods are consistent here. **Questions 22 to 25** require the correct application of the rules of BIDMAS.

Questions 26 and 27 require students to use the idea of equivalent fractions to first order fractions and then find mid-way fractions. **Questions 28 to 30** are problem-solving questions which could be given as extension work.

Exercise 4 looks at converting between decimals and fractions. The first 40 examples are basic practice while **questions 41 to 48** ask students to add, subtract, multiply or divide when the numbers are in different forms. Encourage students to work in whichever form they prefer rather than insisting on one particular approach. **Questions 49 to 52** require students to order numbers using the idea of equivalence.

The exercise finishes with questions where students convert from recurring decimals to fractions. Ensure they are multiplying by the correct power of 10.

Plenary

A decimal/fraction card-matching activity could be given to students to complete in pairs. One pair could then check agreement with another pair and whole-class feedback can be given at the end. An alternative approach would be to give the students a prepared worksheet with lists of decimals and fractions that they have to match.

Lesson 3 – Number facts

Textbook pages 7–10

Expected prior knowledge Students should be familiar with finding factors and multiples of small numbers (see starter activity).

Objectives
E1.1: Identify and use natural numbers, integers, prime numbers, common factors and common multiples, rational and irrational numbers.

Starter
Multiple and factor identification: Ask students to write down seven or eight numbers less than 50. Generate random numbers between 1 and 50 and tell students to cross their chosen numbers off if they are either a factor or a multiple of the random number. The first student to cross all their numbers off wins the exercise.

Lesson commentary
● Prime numbers under 100 can be revised using Eratosthenes' sieve on a 100 number square. An interactive version on a 400 number square can be found at http://www.hbmeyer.de/eratosiv.htm (link correct at time of publication). Printed number squares can be given to the students to help them.

● Discuss the generalised method for testing for 'primeness' (divide by every prime less than the square root of the given number) and ask students to check some examples using standard divisibility rules and a calculator (for bigger numbers).

● In the second part of the lesson (or in the next lesson, if necessary) explain the different types of number which there could be and provide examples (natural numbers, integers, rational numbers). Discuss the meaning of irrational numbers (numbers that cannot be written as a ratio of two integers). Provide examples and ask students to classify a list of numbers provided. (This could be projected onto a screen or given as a worksheet. Examples could also be taken directly from the textbook.)

● Students might be familiar with writing numbers as the product of prime factors but it is worth working through an example and linking this to finding HCFs and LCMs. Students can then practise these skills with additional examples.

Exercise commentary

Exercise 5 looks at classifying primes, listing multiples and factors and finding simple common multiples, as well as prime factorisation and finding HCFs and LCMs. **Question 7** could be used as an extension activity.

Exercise 6 asks students to first classify and then work with rational and irrational numbers. Emphasise the need to give reasons for their answers throughout.

Plenary
Ask students to give examples of numbers with certain features. Examples could include 'a rational number with a denominator of 3', 'an irrational number between 5 and 7' or 'a number with exactly three factors'.

Lesson 4 – Sequences

Textbook pages 10–14

Expected prior knowledge Students should have met sequences beforehand so the emphasis should be on generating nth term formulae.

Objectives

E2.7: Continue a given number sequence. Recognise patterns in sequences including the term-to-term rule and relationships between different sequences. Find the nth term of sequences.

Starter

Simple addition/multiplication exercises – brain training: Pick a start number. Get students to successively add, subtract and multiply the start number. Try 10 operations and ask students to record the final answer.

Lesson commentary

- Students should be able to work out the next few terms of simple sequences already and a quick question and answer session will enable them to practise this. Start with simple examples such as 1, 3, 5, 7, ... and then extend this to include multiplicative sequences (1, 3, 9, ..., for example). Increase the complexity so that they are dealing with quadratic sequences such as 2, 5, 10, 17, ... and if appropriate, Fibonacci-style sequences such as 2, 3, 5, 8, ...

- Ask students to generate their own sequence and challenge a partner to continue it. Emphasise the need for a constant rule between terms. Students could then work in pairs to challenge another pair to spot the rule of a more complicated sequence. For example, one generated by multiplying successive terms by 2 and adding 3 (e.g. 1, 5, 13, 29, ...).

- The main part of the lesson should be about finding and identifying the nth term formulae for sequences. Using a series of examples, encourage students to think of the *common difference* as the multiplier for n and then adjust accordingly. For example, 3, 7, 11, ... will be of the form $4n + ...$, what is the ...? (in this case, '−1').

- Students can be given further examples and asked to practise generating the nth term formulae for a range of linear sequences.

- Sequences which arise out of 'matchstick' patterns or other geometrical arrangements can be introduced. Ask students to identify the pattern and continue it before tabulating and working out the nth term formulae. At this point, they could be encouraged to generate their own patterns and deduce the formulae for them.

- Non-linear patterns can be introduced as appropriate. Students should be encouraged to solve these by comparison to n^2 or n^3 by using a method of subtraction as shown in the example. Exponential sequences should be identified and solved by inspection.

Exercise commentary

Exercise 7 asks students to simply write down the next terms for a variety of sequences and to state the term-to-term rule for some of them. It could be done as a whole-class question and answer exercise.

Exercise 8 looks at the idea of an nth term formula. In **questions 1 and 2**, students match sequences with given formulae while in **questions 3 and 4** they are asked to generate sequences from given formulae.

In Exercise 9, students are expected to work out the nth term formulae for themselves.

Questions 1 to 10 in exercise 10 all relate directly to n^2. **Questions 11 to 13** relate to n^3 and there is a hint to guide students. Likewise, a hint is given for **questions 14 and 15** as these sequences are exponential in nature.

Plenary

Give students a set of sequences and a set of rules. Ask them to match the sequences with the rules and then continue the sequences for two more terms. These could be provided using prepared worksheets, matching cards or projected onto a screen.

Lesson 5 – Rounding and estimation

Textbook pages 14, 18–19

Expected prior knowledge It is expected that students will be familiar with applying standard rounding conventions (see starter activity) but further revision of these methods may be necessary.

Objectives

E1.9: Make estimates of numbers, quantities and lengths, give approximations to specified number of significant figures and decimal places and round off answers to reasonable accuracy in the context of a given problem.

Starter

Give the students a quick question and answer quiz where given numbers are rounded to the nearest 10, one decimal place, etc.

Lesson commentary

- Students should be familiar with the concept of rounding numbers to the nearest ..., a given number of decimal places and a given number of significant figures. Revise these ideas as necessary, either using the exercise in the textbook or through a more extended question and answer session building on from the starter activity.

- Key problems might include truncating large numbers when rounding to a certain number of significant figures (23 450 rounds to 23 000 (2 s.f.) rather than just '23'), and confusion about the place of zeros in both rounding conventions. All zeros are counted when rounding to a number of decimal places, whereas only the zeros which come after the first non-zero digit are counted when rounding to a number of significant figures.

- One key area where rounding is important is in estimating the size of the answers to calculations. Students should be able to make sensible estimates for the value of certain calculations. Provide them with some examples of this type and ask them to round each number to one significant figure before working out the estimate. They can then check the accuracy of their calculations by working out the actual answer on a calculator. As an extension, ask them to also work out the percentage error between their answer and the actual answer.

Exercise commentary

Exercise 11 is a rounding exercise and could be done as a whole-class question and answer session if appropriate.

Exercise 14 is an exercise in estimation. Students should be advised that rounding to one significant figure is the accepted convention when estimating. After completing their estimates, students could check the answers using a calculator.

Plenary

A calculation quiz could be given where students have to estimate the answers to a number of questions, for example 3.13×4.98.

Lesson 6 – *Measurements and bounds*

Textbook pages 15–18

Objectives
E1.10: Give appropriate upper and lower bounds for data given to a specified accuracy. Obtain appropriate upper and lower bounds to solutions of simple problems given data to a specified accuracy.

Starter
Ask students to give examples of numbers which would be rounded to a given degree of accuracy. An example could be 'give numbers which are rounded to 4.3 (to one decimal place)' and expect responses such as 4.27 and 4.32.

Lesson commentary

- Following on from the previous lesson, a second aspect of this work on approximation and estimation is the idea of bounds for measurements. Emphasise that measurements are necessarily rounded to the nearest ... and that the true value could lie within half a unit either side (due to the conventions of rounding). Students should have no problem accepting the lower bound as 0.5 below but they may take some convincing that the upper bound is 0.5 above ('but sir, that value rounds up!'). Discuss the idea that 0.4 can be 'beaten', as can 0.49, as can 0.499, etc. and therefore it makes sense to use 0.5 as the upper bound with the proviso that the actual value is strictly *less than*.

- Provide students with examples of calculations involving bounds and discuss the methods for getting the upper and lower bounds for the resulting sum. Emphasise that for subtraction and division, the lower bound is found by 'lower bound (operation) upper bound' and the upper bound is found by 'upper bound (operation) lower bound'.

Exercise commentary
Exercises 12 and 13 test the students' understanding of bounds in measurement and the questions in exercise 13 deal with calculations involving bounds.

Plenary
In order to effectively assess students' progress, ask them to work out further simple sums involving bounds and write the answers in their exercise books for checking at the end.

Lessons 7 and 8 – Standard form

Textbook pages 19–21

Objectives

E1.7: Use the standard form $A \times 10^n$ where n is a positive or negative integer, and $1 \le A < 10$.

Starter

Ask the students to write down the answers to, or provide via question and answer, a series of 'powers of 10' questions. If you allow the use of calculators, you could include negative powers as well.

Lesson commentary

- Explain that very large numbers or very small numbers which have lots of zeros after or before are quite difficult to manipulate. Provide examples where these numbers might be encountered (space measurements, atomic masses/sizes, etc.) and explain that there is a way that these numbers can be represented to make them more manageable.

- Introduce the idea of 'standard form' through some simple examples, both large and small, and refer back to the starter activity for powers of 10. Ensure students are comfortable with the value of A being between 1 and 10 (could be decimal but not *equal* to 10). Explain how the power is determined (move the *digits* until there is single digit in front of the decimal point).

- Give them a series of examples of numbers written in standard form and ask them to convert them to normal numbers. Likewise, get them to work the other way. Address any problems with this conversion process as you go along and then explain how you use a calculator to work with numbers given in standard form (usually using the 'EXP' key).

- Students can then have a go at some more problem-solving type questions which use numbers given in standard form before introducing calculations involving numbers in standard form. Model multiplication and division examples and discuss the use of a calculator in this context.

- The second lesson can be used for further consolidation and/or problem-solving with numbers given in standard form.

Plenary

In order to check the students' understanding of this topic, ask them to do further conversions and/or calculations using standard form. This could take the form of a quiz or short test.

Lesson 9 – *Ratio*

Textbook pages 21–23

Expected prior knowledge At this level, students should be familiar with using simple ratios so this lesson provides a good opportunity to revise some of the key applications of ratio.

Objectives

E1.11: Demonstrate an understanding of ratio and proportion. Increase and decrease a quantity by a given ratio.

Starter

Give the students questions such as 'divide 100 by 10 and multiply the answer by 3', 'divide 360 by 12 and multiply the answer by 7', etc. Increase the complexity of the questions as appropriate.

Lesson commentary

- Students should be familiar with using simple ratios to divide quantities up but applying them to more complicated problems may need careful revision. Provide the class with some simple examples involving mixing paint, cordial and water mixes, etc. and ask them to respond by writing the answers in their exercise books or by whole class question and answer. How much red paint would be needed to make 15 litres of pink paint if red and white are mixed in the ratio 2 : 3, for example?

- Provide further examples where one of the sub-quantities are given and students are expected to work out the original amount that was divided up. For example, if John receives $70 and the money was divided in the ratio 7 : 3, how much money was originally divided up?

- Students should also be given examples where ratios are to be simplified into the form $1 : n$ (or $n : 1$). Problems of this type can be presented as a simple exercise in division and may require the use of a calculator.

- Allow students time to consolidate this work through the use of textbook exercises or other similar sets of questions.

Exercise commentary

Exercise 17 tests students on basic ratio techniques such as dividing into and writing ratios in the form $1 : n$. **Questions 26 to 30** are more difficult and could be used as extension questions.

Plenary

To assess the students at the end of the lesson, provide them with a summary question and answer session where they solve a variety of ratio problems quickly. Allow the use of calculators if appropriate.

Lesson 10 – *Direct proportion*

Textbook pages 23–24

Objectives
E1.11: Demonstrate an understanding of proportion.

Starter
What is one of ...? Give the students a scenario, for example 8 apples cost 96 cents, and ask them to work out what one apple would cost. Repeat for different situations and include questions which require the use of a calculator.

Lesson commentary
- The use of the 'unitary' method is a standard approach for direct proportion. Ensure students are happy with the principle used in the starter and then discuss the unitary method more fully. What would happen if you were given the 8-apple scenario but then asked for the price of five rather than one? Going via one apple is clearly the easy way to proceed.

- Provide students with further examples of direct proportion where the unitary method can be used. Students can work together on problems of this type, if appropriate.

- Ask students to think of examples of direct proportion that they may encounter in the real-world. Purchasing items from a store such as fruit and vegetables is a common example but examples could also include buying gas and currency exchanges.

Exercise commentary
Exercise 18 contains a mixture of questions on both direct and inverse proportion. **Questions 1, 2, 4, 5, 6, 7, 9, 13, 14 and 15** are good examples of direct proportion. Exercises 19 and 20 look at the application of proportion to exchange rates and map scales. These could be introduced at this point or set aside for future consolidation work.

Plenary
Give the students a situation such as '9 pencils cost $1.08'. Ask them to quickly write down the answers to a number of follow-up questions such as 'How much would 6 cost?', 'How much would 10 cost?' and 'How much would 100 cost?'

Lessons 11 and 12 – Inverse proportion

Textbook pages 23–27

Objectives

E1.11: Demonstrate an understanding of proportion.

E1.15: Calculate using money and convert from one currency to another.

Starter

Start with a number, say 100, expressed as 100×1. Ask students to write down the multiplication sum that would make 100 if the '100' part of the multiplication was successively divided by two. Challenge the students to take the sequence as far as they can. The first few terms would be 50×2, 25×4, 12.5×8, etc.

Lesson commentary

- Students often find problems of inverse proportion difficult to understand. A common problem is with the *recognition* of when two quantities are inversely related so introduce the topic through a range of examples including men digging holes and the relationship between speed and time. Encourage active discussion of the problems as students work on them.

- Ask students if they can think of situations in the real-world that show an inversely proportional relationship. Further examples could include the relationship between density and volume or decorators painting a house.

- Allow sufficient time for students to consolidate this work and, if appropriate, begin to mix up the types of question to include examples of direct proportion as well. This will encourage students to think carefully about the problems rather than just assume they are of a single type.

- The second lesson can be used for consolidating this work on proportion further and for looking at exchange rate applications and map scales (see exercise commentary).

Exercise commentary

Exercise 18 contains a mixture of questions on both direct and inverse proportion. **Questions 3, 8, 10, 11, 12 and 18** are good examples of inverse proportion.

Exercises 19 and 20 look at the application of proportion to exchange rates and map scales. These can be taught separately or incorporated within the lessons on this topic.

Exercise 21 extends the idea of map scales to look at area conversions and could be reserved as an extension activity for more able students.

Plenary

In order to test the students' understanding of the work on proportion, a mixed question and answer activity can be carried out. Provide the students with a number of relatively straightforward examples of both direct and inverse proportion and ask them to either write the answers down in their exercise books or respond orally.

Lessons 13, 14 and 15 – *Percentages*

Textbook pages 28–32

Expected prior knowledge Students should be familiar with the equivalence between decimals and percentages (see starter activity) and they should also be able to calculate simple percentages of amounts. This series of lessons provides an opportunity to revise these ideas before developing them further.

> ## Objectives
> E1.12: Calculate a given percentage of a quantity. Express one quantity as a percentage of another. Calculate percentage increase or decrease. Carry out calculations involving reverse percentages.

Starter

Fraction, decimal and percentage conversions: Give the students a series of questions asking them to convert between commonly used fractions, decimals and percentages.

Lesson commentary

- Discuss the possible calculator and non-calculator methods for working out the percentage of an amount. These could include breaking the percentage down into 10%, 5%, 1%, etc. or via the unitary method of dividing by 100. Discuss the idea of a multiplier (the decimal equivalent of the percentage: 20% = 0.2, for example).

- Discuss the link to percentage increase and decrease. 'If I can work out a percentage *of* an amount, then I can just add it or subtract it to work out a percentage increase or decrease.'

- Using a simple example, demonstrate the technique for finding the percentage increase (actual increase divided by original amount, multiplied by 100). Students will have to do this without a calculator for simple examples so encourage them to efficiently manipulate the fractions. Students will use a calculator for harder examples.

- Ask the students how they might modify the method to calculate percentage *decrease*. Provide them with a few examples to try (both with and without a calculator) and then they can do further practice as necessary. Link to profit and word-based problems as appropriate.

Exercise commentary

The questions in exercise 22 could be used as a starter activity or done via whole-class question and answer.

Exercises 23 and 24 deal with the full range of percentage calculations and suitable questions can be selected from either to consolidate the techniques at different points throughout this series of lessons. The key ideas behind each one are the same and students should be encouraged to think carefully about each question rather than making assumptions that a block of questions are all of the same type. The later questions in both exercises could be used to extend stronger students.

- Ask students to come up with real-life examples where percentage increase and decrease are used (in-store sales, tax, service charges, etc.) They could set their own problems and challenge a partner to work out the new values.

- Link all of this into the idea of the multiplier and percentage change. Establish the basic rule 'original value × percentage multiplier = new value'. Model examples where the new value is to be found and also where the original value is to be found (reverse percentage change). A common mistake when finding the original value is to find the percentage of the new value and subtract it. For example, 'x is increased by 15% and you get y, therefore to get back to x, find 15% of y and subtract it'. Emphasise that x is always '100%' and if students continue to get these types of problems wrong, encourage them to check their solutions afterwards.

- There are a number of key percentage calculations that students are expected to be able to carry out. These include calculating the percentage of an amount, increasing and decreasing by a percentage, calculating percentage increase and decrease and calculating using reverse percentage change. This series of lessons is designed to cover all of these and provide students with the opportunity to practise them. It is at the discretion of the teacher the order in which they are taught but the lesson commentary provides a *suggested* order. Additional starter and plenary activities can be incorporated as appropriate.

Plenary

Provide students with questions and worked solutions which contain mistakes and ask them to study the solutions and identify the mistakes, correcting them as appropriate. This could be done as a paired activity, and be used to address some of the key misconceptions noted above.

Lesson 16 – *Interest and tax*

Textbook pages 32–35

Objectives

E1.16: Use given data to solve problems on personal and household finance involving earnings, simple interest and compound interest.

Starter

Ask students to use a calculator and record their answers on a mini-whiteboard or in their exercise book, provide them quick 'multiplier' questions. Examples could include 25 × 1.05 and 124 × 1.12.

Lesson commentary

- Discuss where students might encounter the concept of interest (bank loans, savings, mortgages, etc.) and set the problem 'would you rather have a rate of 5% on your savings and get the money earned back every year, or a rate of 4% which is automatically reinvested?' 'How much would you have in each case after 1 year, 2 years, etc?'

- Discuss the link between percentage multipliers and the method for working out interest. Ask students to use a table to generate the growth of money in an account under both interest schemes. At what point does the 4% compound interest overtake the 5% simple interest?

- Students could then be given further examples where they have to work with both simple and compound interest.

- At Extended level, students should be able to work with both simple and compound interest in the context of a single lesson but if necessary, the lesson could be divided so that each type of interest is dealt with separately.

- Students should be familiar with using the compound interest formula and they need to learn it. Encourage them to use it when possible and get into the habit of inputting the calculation into their calculators in one go.

- The short exercise on tax which follows the work on interest could be set as homework for the students. They should follow the example and then try to solve the problems in the exercise. Alternatively, this work can be introduced as part of the second lesson on interest.

Exercise commentary

Exercise 25 contains examples on simple interest and **question 3** requires students to work backwards. Exercise 26 contains examples on compound interest. but **question 7** involves *depreciation* and it should be checked that students are deducting the amount at the end of each year. **Questions 8, 9 and 10** could be used as extension questions while **question 11** links in with the plenary activity.

In exercise 27, all questions follow the structure of the income tax example but **questions 1 and 2** have only one basic tax rate.

Plenary

Give the students three minutes to work out whether they would rather have 10% simple interest or 7% compound interest if they invested $100 over 5 years (10% simple interest is $150, 7% compound interest is $140.26 so simple interest is better). As an extension task, you could ask students to work out after how many years they would have more under the compound interest scheme.

Lesson 17 – Speed, distance and time, other rates

Textbook pages 36–39

Objectives
E1.11: Calculate average speed. Use common measures of rate.

Starter
Give students a quick question and answer activity testing simple division. Examples could include 'What is $120 \div 3$?' and 'What is $240 \div 6$?'

Lesson commentary

- Students should be familiar with idea of average speed but may not have dealt with it formally. Discuss how you might interpret speed/ distance/time information when given it for a scenario involving travelling to a nearby city. For example, 'If I start at home and travel a distance of 120 kilometres, how long would it take me at an average speed of 50 kilometres per hour?' Discuss the idea of using division to solve such a problem.

- What if I know the speed and the time? How could I use this information to work out the total distance of my journey?

- Develop the idea of the speed/distance/time 'triangle' and explain the 'cover up' method for working out the unknown quantity.

- Provide students with examples to practise (encourage collaborative working) and then consider an example where the information is given in metres and seconds and discuss how such a speed might be calculated in kilometres per hour. Is it easier to convert the units first or convert the final answer? Allow students flexibility to choose their own method.

- Explain that there are other rates that students might encounter such as litres per minute when filling a bath with water. Focus on the word *per* as indicating division of the two quantities: amount of water *divided* by the time gives the rate.

Exercise commentary

The questions in exercise 28 fall into two categories: finding speed, distance or time from the information given (using the 'triangle') and converting between units of speed. **Questions 5, 6 and 7** require students to interpret multi-stage problems and work out overall average speed. **Questions 8 onwards** could be set aside for extending more able students as they are much more involved.

In **question 1** in exercise 29 students are finding the rate, whereas in **question 2** they have to find the time taken so a rearrangement of the 'formula' will be required. **Questions 3 to 6** are of a problem-solving nature. Encourage students to write down all the information they know before proceeding.

Plenary
Ask students to write down approximately how far they live from school and the approximate time it takes them to get to school in the morning. Discuss modes of transport and get them to work out an estimate of the speed at which they travel to school.

Exercises 30, 31 and 32 on pages 40–42 are mixed exercises testing students' practical application of number skills and could be set as homework or used in a consolidation lesson.

Lesson 18 – *Using a calculator*

Textbook pages 42–47

Objectives
E1.13: Use a calculator efficiently. Apply appropriate checks of accuracy.

Starter
Give the students a start number and then give them a series of operations to perform on a calculator. Make sure that they press the 'equals' key after each one. Ask for a volunteer to give their final answer and check that this matches the final answer of the rest of the class.

Lesson commentary

- Since there are many different makes and models of calculator on the market, it is difficult to *teach* calculator skills but there are several useful hints and tips that you can give to students to help them use their calculators efficiently.

- Make sure that they know where the key buttons are such as 'x^2' and '$\sqrt{\ }$'. Also make sure they can find the 'power' button (usually a '$\wedge$' or 'x^y').

- Make sure that they are confident in their understanding of the rules of BIDMAS and that they take these into account when keying in sums to the calculator. As a general rule, 'if in doubt put some brackets in' and 'always bracket numerators and denominators when dealing with fractions'.

- Make sure they are aware of the function of the 'ANS' key when doing multi-stage calculations (the 'memory' functions are not as useful as they once were since the introduction of DAL (Direct Algebraic Logic) calculators and the 'invention' of the 'ANS' key).

- Always put negative numbers in brackets (and use the '(−)' key rather than 'take away').

- Finally, encourage them to experiment and have the answers available to help them to check that what they are keying in correctly gives the answer.

Exercise commentary
Exercises provide plenty of calculator practice. Ensure students take into account all of the advice given and it is also worth emphasising the need to have an idea of the *size* of the answer beforehand.

Exercise 35 emphasises the need to check calculations for sense as well as the usefulness of estimating the answers beforehand.

Plenary
Further calculations with the answers written in exercise books could be used or students could challenge each other to come up with an 'interesting' sum that has a particular given answer.

The revision and examination-style exercises can be used for further practice as appropriate.

CHAPTER 2
ALGEBRA 1

Lesson 1 – Directed numbers

Textbook pages 56–59

Expected prior knowledge Students should be familiar with the arithmetic of negative numbers so this lesson provides a good opportunity to revise these ideas.

Objectives
E1.4: Use directed numbers in practical situations.

E2.2: Manipulate directed numbers.

Starter
Number line trail: Give the students a starting number, for example 3, and then ask them to work out the finishing number when a series of addition and subtraction sums are applied successively to it. For example: $+3-7+5-6+2-5+1 (=-7)$. Repeat the process for a different starting number and a different string of calculations. Include multiplication and division if appropriate.

Lesson commentary
- Students may need reminding of the key rules when adding/subtracting or multiplying/dividing negatives. Ensure they do not get the rules mixed up (adding two negative numbers *does not* give a positive answer, for example). Demonstrate examples of each type of calculation if necessary.

- Put negative numbers in context by discussing real-life situations such as temperature, depths of the ocean, etc. and ask students to try answering some examples.

- Thoroughly revise calculating with directed numbers through appropriate examples and question practice. There are lots of questions in the exercises for this lesson but able students will not need to do all of them. Use an appropriate approach such as 'do just the odd numbers' or 'just do the questions in the first column' to ensure students answer enough questions to consolidate the topic effectively.

Plenary
Quick negative number arithmetic: This could be done either using mini-whiteboards or as a short test to assess understanding.

The support worksheet for Chapter 2 may be useful to support lessons 4, 12 and 13.

The challenge worksheet for Chapter 2 may be useful to support lesson 28.

Exercise commentary
Exercise 1 involves using negative numbers in context (temperature, etc.) and these examples can be incorporated into the main body of the lesson if appropriate.

Exercises 2 and 3 provide routine practice in applying the rules for directed number arithmetic and can be used for revision and consolidation.

Exercise 4 is a mixed exercise and able students should be encouraged to complete this rather than exercises 2 and 3.

Lessons 2 and 3 – *Formulae*

Textbook pages 59–63

Objectives
E2.1: Use letters to express generalised numbers and express basic arithmetic processes algebraically. Substitute numbers for words and letters in complicated formulae.

Starter
Provide students with a simple algebraic expression ($2a + b$, for example) and then generate values for the unknowns (using different polyhedral dice, for example). Ask students to apply the substitutions and write the answers in their exercise books.

Lesson commentary

- Explain that students will be substituting numbers (including negatives) into a succession of formulae, some of which are simply abstract algebraic expressions and some of which are real-life formulae in context. Provide them with some examples (harder than the starter expression) and ask them to work together to check that they are correctly substituting the numbers and arriving at the correct answer.

- Increase the complexity of the expressions, introduce negative numbers and possibly numbers given in standard form (see exercise 5 question 6).

- Ask students to produce their own expressions (sensible ones) and challenge a partner to substitute a series of numbers into them.

- Students could be given a further activity where a series of expressions are given along with the answers when a certain set of numbers are substituted in. They should attempt to match the expressions with the values.

- The second lesson will enable students to have sufficient time to consolidate this work fully. There are a lot of questions in the exercises for this section so encourage students to attempt a variety of questions at different levels of complexity, rather than just doing 20 or 30 of the same type.

Exercise commentary
Exercise 5 requires students to substitute numbers into formulae in context and could be included in the main body of the lesson. **Questions 8 to 11** develop the idea of writing simple formulae from information given.

Exercises 6, 7 and 8 provide many examples of abstract expressions and negative substitutions and these questions can be used as appropriate. The examples in exercise 6 are standard linear expressions and most students should be able to attempt these. The examples in exercise 7 incorporate powers and brackets and some students may need guidance on these. Emphasise that a negative number squared is positive.

Exercise 8 can be used as a challenge exercise for more able students as the expressions are significantly more complex than in the previous two exercises.

Plenary
Order the expressions. Provide students with four expressions and then ask them to order them by size (increasing) upon substituting a given set of numbers. Repeat for different sets of numbers.

Lesson 4 – *Brackets and simplifying*

Textbook pages 64–65

Objectives
E2.2: Expand products of algebraic expressions.

Starter
Multiplication of algebraic terms: Ask the students a series of questions such as 'What is $3 \times 2x$?', 'What is '$4 \times 5y$?' and 'What is $7 \times 2z$?'

Lesson commentary

- Most students will be familiar with expanding simple brackets but they will certainly benefit from further practice. Demonstrate a number of examples, emphasising the need to multiply both of the terms inside the bracket by the term outside. Encourage them to draw arrows over the top of the bracket to indicate these two actions. Provide further examples for practice and encourage them to check their answers with a partner. Provide examples with negative terms outside and also examples with more than just a single bracket.

- Move on to collecting like terms and introduce brackets in this context as well. Students should be able to combine the two skills and include squares and negative terms.

- Students can then be asked to complete a series of questions from the exercise in order to consolidate this work fully.

Exercise commentary
Exercise 9 provides lots of examples of algebraic simplification. The first 24 questions are on simplification only while **questions 25 to 48** introduce brackets in this context. More able students could begin on these directly.

Plenary
Provide students with a card-matching activity where they are given a series of expressions and their simplified equivalents. Ask them to work in pairs to match them up. Alternatively, ask the students further questions on brackets and simplifying and instruct them to write the answers in their exercise books or respond orally.

Lessons 5 and 6 – *Two brackets*

Textbook pages 65–67

Objectives
E2.2: Expand products of algebraic expressions.

Starter
'Right or wrong?' Provide students with a number of pairs of expressions and ask them to say whether they are equivalent or not. For example, 'Is $3(x + y)$ equivalent to $3x + 3y$?' (yes) and 'Is $-2(2x + 3y)$ equivalent to $-4x + 6y$?' (no).

Lesson commentary
- There are many methods available to students when expanding two brackets. The method used in the textbook is one alternative but there is also the grid method and FOIL (first, outside, inside, last). Demonstrate some examples using the different methods and ask students to use the one that they prefer for further examples. Students could then be grouped according to this choice and asked to work together to solve further expansion problems.

- Discuss the idea of a perfect square (a repeated bracket) and ensure that students write these types of problem as a product of two brackets before removing them. The common mistake when expanding perfect squares is to simply square the first term and square the second rather than considering them as the product of two separate brackets.

- Standard pairs of brackets and perfect squares could be dealt with in two separate lessons if appropriate. Alternatively, both types could be introduced in a single lesson and the second lesson used to provide time for consolidating this work on expanding pairs of brackets.

Exercise commentary
Exercise 10 provides routine practice in expanding pairs of brackets. From **question 21 onwards**, there is a third component in front of the two brackets which needs to be taken into account at the end. Discourage 'short-cutting' of steps.

Perfect squares are introduced in exercise 11. **Questions 15 onwards** and all of exercise 12 require two expansions followed by simplification of terms. These questions could be set aside as extension problems for more able students.

Plenary
Provide students with a number of worked solutions to expansion problems which contain one or more errors. Ask them to work in pairs to identify the mistakes and provide a correct solution to the problem. Alternatively, give the students further expansion examples and ask them to write the answers in their exercise books or respond orally.

Lessons 7, 8 and 9 – *Linear equations*

Textbook pages 67–72

Expected prior knowledge Students should be familiar with solving simple linear equations. These lessons can consolidate and develop this to include equations with a more complex structure.

Objectives
E2.5: Derive and solve simple linear equations in one unknown.

Starter
'Think of a number': Tell the students that you are thinking of a number and that when you, say: 'add 2 and double it', you get, for example, 10. What number did you think of? Responses can be via mini-whiteboards, written into exercise books or given orally by a member of the class.

Lesson commentary

- Students should be able to solve simple equations of the form $ax + b = c$. An introductory activity can be used to establish this by giving a sequence of examples and asking students to work out the values of the unknowns and write down their responses. Students may feel less confident with 'take away' and 'divide' and negative or fractional answers can also confuse some students who think the answer must always be a positive integer. Increase the complexity of the examples as appropriate to test this.

- Move on to examples where the unknown is on both sides. Ask students to test the validity of certain answers using 'true' or 'false'. For example, 'Is $x = 4$ a solution to $3x + 2 = 2x + 6$?' Discuss the need to 'balance' the equation by always doing the same thing to both sides and encourage the collection of the unknowns on the side which has the most. Provide examples as appropriate and then allow students time to consolidate. Again, negatives are likely to cause confusion so encourage students to remove negatives where possible by adding terms to both sides.

- In addition to equations where x appears once or on both sides of an equation, other examples involving brackets, pairs of brackets and equations where x is in the denominator need to be covered. A worked example approach is generally the best way to ensure students develop the methods correctly. Encourage students to develop individual strategies for each type of equation such as removing negatives and fractions at an early stage and simplifying where possible.

Exercise commentary

Questions 1 to 31 in exercise 13 have a single unknown while the remaining questions are a mixture of two unknowns, fractions, decimals and negatives. More able students should be able to work quickly through some simple examples before moving on to the later questions, while less able students will benefit from carefully working through the first part of the exercise.

Exercise 14 introduces brackets and the questions increase in complexity as they go on. Again, encourage more able students to move on to the later questions.

Pairs of brackets (and perfect squares) are the focus of exercise 15 and students should be encouraged to simplify whenever possible.

In exercise 16, the unknown is often in the denominator and all of the questions involve fractions. Encourage students to remove fractions at an early stage and collect like terms. More able students could be instructed to tackle the later problems in order to provide appropriate challenge.

Plenary
Provide worked examples of equation-solving which contain errors. Get the students to work in pairs to decipher the solution, find any mistakes and provide corrected solutions. Differentiate the questions as appropriate.

Lessons 10 and 11 – *Problem-solving using linear equations*

Textbook pages 72–77

Objectives
E2.5: Derive and solve simple linear equations in one unknown.

Starter
Ask students to write down five numbers between 1 and 20. Roll a die and then perform a simple operation on the number, say, double it or add 7. If the students get the answer, they can cross it off their list.

Lesson commentary
- The emphasis here is to interpret the information from a word problem and create an equation which can then be solved by standard methods. Start by providing simple word problems along the lines of 'think of a number' and get the students to write down the equation that results. Students may not need to formally solve the equation but they should be encouraged to check that their intuitive solution and their equation do give the same answer.

- Provide students, possibly working in pairs or small groups, with practical problems involving the area and perimeter of rectangles, angle problems, consecutive number sums, etc. These could be provided on prepared worksheets or taken from the exercises. Ask them to write down the equations needed to solve the problems and then to find the value of the unknown. The task can be made different by providing an appropriate selection of questions for each student, pair, or group. Students could then be asked to make up their own problems of this type and challenge others. Emphasise the need to work out their own solutions first to make sure that they are consistent.

Exercise commentary
Exercise 17 is a series of typical examples which increase in complexity as they go on. Some students may have difficulty in identifying the 'unknown' and collaborative working on questions which provide appropriate challenge should be encouraged.

Exercise 18 is more difficult and these questions could be reserved for more able students working to provide model solutions.

Plenary
Provide students with further word problems (read aloud to encourage listening skills) and challenge them to solve the problems in the quickest time.

Lesson 12 – *Solving simultaneous equations by substitution*

Textbook pages 77–78

Objectives
E2.5: Derive and solve simultaneous linear equations in two unknowns.

Starter
Algebraic rearrangement: Give the students an example such as
$x + 2y = 4$ and ask them to make x the subject. Then ask them to then
make y the subject. Change the example and repeat. This activity will
enable students to get some initial practice in rearranging formulae.

Lesson commentary
- The topic of simultaneous equations could be introduced using a
  classic example of apples and bananas (or equivalent). Suggest the
  problem that you want to find the cost of one apple and one banana
  but that when you went into the fruit and vegetable shop, they sold you
  two apples and three bananas for 64 cents and the next time you went
  in they sold you four apples and 2 bananas for 80 cents. How much
  is one apple and one banana? (Apple: 14 cents, banana: 12 cents.)

- One method of solving simultaneous equations is the substitution
  method where one of the equations is given in (or rearranged into)
  the form '$y = ...$' or '$x = ...$'. Students need to be comfortable with
  algebraic manipulation to solve simultaneous equations in this way
  and the starter activity is designed to check this understanding.
  If students struggle with the starter activity, further practice in
  rearranging formulae of the type should be given.

- Demonstrate a solution using the method of substitution and make
  sure that the students can follow the algebraic steps. Ask them to
  complete an example of their own with guidance (possibly working
  together, if appropriate) and then they can practise further examples.

- This approach could be applied to examples from exercise 21
  where simultaneous equations are required to solve a range of
  problems. Students could be directed to these if they quickly
  understand the necessary skills.

Exercise commentary
Exercise 19 contains several examples which students need to solve by substitution and these can be used as appropriate for consolidation and further practice.

Plenary
Give the students a further example of a pair of simultaneous equations
to be solved by substitution. Allow them two minutes to work out the
solution before writing it down. This can be repeated with different
examples as time allows.

Lesson 13 – *Solving simultaneous equations by elimination*

Textbook pages 79–81

Objectives
E2.5: Derive and solve simultaneous linear equations in two unknowns.

Starter
Provide students with an algebraic expression ($2x + y$ or similar) and then provide numbers for the unknowns and ask students to quickly substitute these in to work out the value of the expression. Change the expression and extend to include negative numbers or squared terms as appropriate.

Lesson commentary
- The elimination method is the most common method for solving simultaneous equations. Use the example of apples and bananas from the previous lesson or another suitable example to demonstrate the solution method including the multiplication of the equations to make the (second) term the same. Use the reminder 'Same Sign Subtract (SSS)' to indicate the next step in the solution and then solve fully.

- Some students may need additional examples which do not require the multiplication of equations in order to develop their understanding of the basic method. These can be provided as appropriate.

- Students can then work through further examples together (with guidance if appropriate) and then practise solving simultaneous equations by elimination with more questions.

- Problem-solving using simultaneous equations can be worked into the lesson as appropriate or looked at separately through further examples.

Exercise commentary
Exercise 20 is inserted to provide practice in manipulating negative numbers before students proceed to the examples in exercise 21 which require the use of the elimination method of solution. **Questions 19 to 21** require initial manipulation and could therefore be omitted or set aside for extension work. **Questions 25 to 30** involve fractions and decimals and could likewise be set aside for more able students.

Plenary
Give the students a further example of a pair of simultaneous equations to be solved by elimination. Allow them two minutes to work out the solution before writing it down. This can be repeated with different examples as time allows.

Lesson 14 – *Problem-solving using simultaneous equations*

Textbook pages 81–84

Objectives
E2.5: Derive and solve simultaneous linear equations in two unknowns.

Starter
Sum and product: Ask students to work out the two numbers when you tell them that the sum is 5 and the product 6 (2 and 3). Repeat for different sums and products.

Lesson commentary

- The previous two lessons have introduced students to the two standard methods for solving simultaneous equations and a quick summary of these is likely to be useful. Some students may also have had time to work on some problem-solving type questions as well.

- By selecting a range of examples, either from the textbook or elsewhere, demonstrate the formulation of the simultaneous equations from the information contained in the question. Emphasise that it is important to get this first action right because then the solution of the pair of equations can be solved by using one of the two standard methods.

- Discuss further examples as appropriate and then allow the students time to practise forming and solving simultaneous equations by completing further examples of their own. Encourage students to check they have correct equations before solving.

Exercise commentary
Problem-solving using simultaneous equations is tested in exercise 22. Students may need help with initially modelling the problem; encourage them to use the method of their choice to find the solutions. **Questions 27 to 30** have three unknowns; these could be given to more able students as extension problems.

Plenary
Give the students a further example and ask them to work out the solution using the method of their choice. They could also be encouraged to set their own problems and challenge a partner to solve them.

Lesson 15 – *Factorising*

Textbook pages 84–85

Objectives
E2.2: Use brackets and extract common factors. Expand products of algebraic expressions. Factorise where possible expressions of the form: $ax + bx + kay + kby$, $a^2x^2 - b^2y^2$, $a^2 + 2ab + b^2$, $ax^2 + bx + c$.

Starter
Provide students with a list of algebraic expressions and ask them to write down the *highest common factor* of certain pairs of expressions. A good set of expressions might be $3x$, $6x$, $3y$, $2y$, x^2, y^2, $2x^2$ and $3y^2$. Include pairs where 'no common factor' is the answer.

Lesson commentary

- Explain that factorisation is the reverse process of taking out brackets. Explain that we are looking for 'how the expression started out' before the brackets were removed. Encourage the identification of the *highest common factor* from two (and eventually three) terms before writing this outside of the brackets. The terms inside the brackets can now be found by dividing each term by this highest common factor. Provide examples before letting the students practise some for themselves. Encourage them to check their answers.

- The factorisation of four terms into two brackets often causes problems for all but the most able students and could be avoided, if appropriate. If this technique *is* introduced, encourage students to follow the examples carefully before attempting to write their own list of instructions for performing this kind of factorisation. Encourage careful thought about the process and clear communication in written instructions. Students may need to consolidate this work more thoroughly than other topics in this chapter and further time could be allowed if necessary.

Exercise commentary
The examples in exercise 23 are routine practice questions where pairs of algebraic terms are to be factorised. **Questions 31 and 32** include three terms and could be used as extension questions.

Four-term factorisation is covered in exercise 24 and students may need clear modelling of this before proceeding.

Plenary
Ask students to write down the answers to several more factorisation questions. Keep them relatively simple and encourage speed of response.

Lessons 16 and 17 – *Quadratic expressions 1*

Textbook pages 85–86

Objectives

E2.2: Use brackets and extract common factors. Expand products of algebraic expressions. Factorise where possible expressions of the form: $ax + bx + kay + kby$, $a^2x^2 - b^2y^2$, $a^2 + 2ab + b^2$, $ax^2 + bx + c$.

Starter

Provide students with pairs of brackets to expand. Encourage them to work quickly and write the answers in their exercise books for checking at the end.

Lesson commentary

- Use examples from the starter activity to introduce the idea of 'reversing the process' and factorising quadratic expressions. Discuss the significance of the two numbers 'b' and 'c' as the sum and the product of the two numbers in the brackets and explain that when we factorise expressions of this type, we try to identify the bracket numbers using factor-pairs of c.

- Students could then be encouraged to work together to solve a range of examples with both numbers positive, both numbers negative and one positive, one negative.

- Discuss examples which have 'no middle term' ($b = 0$) as an introduction to the 'difference of two squares' examples in exercise 27.

- The second lesson will enable students to have sufficient time to fully consolidate this work before moving on to harder examples in the next section.

Exercise commentary

The questions in exercise 25 are routine practice examples of factorising quadratic expressions.

Questions 25 to 27 are slightly harder as the value of c is large and these could be used as extension questions for more able students.

Questions 28 to 30 lead into the work on 'difference of two squares' covered in a later lesson.

Plenary

Provide students with a series of examples where some are correct and some incorrect. Ask them to decide which are correct or incorrect either by expansion of the brackets or factorisation of the quadratic expression.

Alternatively, provide students with a set of dominos containing expanded and factorised expressions and ask them to join them up correctly.

Lessons 18 and 19 – *Quadratic expressions 2*

Textbook page 86

Objectives

E2.2: Use brackets and extract common factors. Expand products of algebraic expressions. Factorise where possible expressions of the form: $ax + bx + kay + kby$, $a^2x^2 - b^2y^2$, $a^2 + 2ab + b^2$, $ax^2 + bx + c$.

Starter

Provide students with pairs of brackets to expand which include say '$2x$' or '$3x$' in one of them. Encourage them to work quickly and display their answers on mini-whiteboards or in their exercise books.

Lesson commentary

- Students often find the method of factorising general quadratic expressions of the form $ax^2 + bx + c$ difficult to follow and consolidate. The best approach is to demonstrate examples and encourage a standard layout for the solution. A step-by-step approach could also be used in this context. Comparing the bracketed form and the expanded form will enable students to 'see' where the numbers come from and how they relate to each other. Encourage them to work together when solving problems of their own and check their solutions by re-expanding the brackets.

- An important check is that their bracket in the 'factorise in pairs' step is the same for both pairs.

- The second lesson will enable students to have sufficient time to fully consolidate this work before moving on to the next section.

Exercise commentary

All of the questions in exercise 26 are routine practice examples of this technique. **Questions 24 to 28** are harder because the value of 'ac' is larger and so it is more difficult to find factors. **Questions 29 and 30** lead into the work on 'difference of two squares' covered in the next lesson.

Plenary

A card-matching activity is ideal here to test understanding. Students could work in pairs to match the cards by equating quadratic expressions and their bracketed equivalents. Alternatively, students can be provided with further examples and work out the factorisation in their exercise books before checking the answers at the end.

Lesson 20 – *Difference of two squares*

Textbook page 87

Objectives

E2.2: Use brackets and extract common factors. Expand products of algebraic expressions. Factorise where possible expressions of the form: $ax + bx + kay + kby$, $a^2x^2 - b^2y^2$, $a^2 + 2ab + b^2$, $ax^2 + bx + c$.

Starter

Ask students to write down the square root of various numbers and algebraic expressions such as x^2, $9y^2$, 25, $\frac{1}{4}$, $16a^2$ and $\frac{x^2}{4}$.

Lesson commentary

- Discuss the special case of a quadratic expression which has 'no middle term'. Reference can be made back to the examples in the previous lessons as an introduction; encourage students to recognise the general form of these expressions:
  $x^2 - y^2 = (x - y)(x + y)$.

- Provide students with several examples using 'non-standard' letters and more complicated expressions for 'x' and 'y'. Link back to the starter activity when deducing the square roots of 'x' and 'y' and allow students time to practise further examples.

- If appropriate, challenge students through the use of examples which require an initial 'simple' factorisation before using the 'difference of two squares' method and then challenge them to solve seemingly difficult arithmetical calculations using this method (examples could be taken from exercise 26, question 29 onwards).

Exercise commentary

Exercise 27 provides students with some routine practice examples which also include fractions. From **question 17 onwards,** initial 'simple' factorisation is necessary to get at the 'difference of two squares' form and **questions 29 to 40** use the idea of 'difference of two squares' to solve arithmetical problems.

Plenary

This is an ideal opportunity to test all of the techniques from this lesson and the previous ones on factorisation. A mixed exercise, or short test could be used for this purpose.

Lessons 21 and 22 – *Solving quadratic equations by factorisation*

Textbook pages 87–89

Objectives
E2.5: Derive and solve quadratic equations by factorisation.

Starter
Factor pairs: Ask students to write down all of the factor pairs of given numbers (10, 16, 12, etc.).

Lesson commentary

- Students could be invited to solve a given quadratic equation (with integer solutions) by trial and improvement. Get them to substitute various numbers into the equation and find one which gives the answer zero. They could then be invited to find a *second* solution to the equation. Explain that quadratic equations will normally have two solutions (reference to the graph could be made at this point) and that there are several methods available for finding the solutions.

- Refer to the methods for factorising quadratic expressions encountered in previous lessons and, using the introductory example(s), demonstrate that the factorised form of the expression provides the solutions to the equation when each of the brackets is set equal to zero.

- Further examples can be demonstrated as appropriate including general quadratics ($a \neq 1$) and special cases such as the difference of two squares and ones with no c term.

- The second lesson will provide students with more time to consolidate this work on solving quadratic equations by factorisation.

Exercise commentary
Exercises 28 and 29 provide lots of basic practice in solving quadratic equations through factorisation. Be selective in the questions given to certain groups in order to provide appropriate challenge and differentiation. For example, weaker students may be instructed to concentrate on examples where $a = 1$ while more able students could be directed to 'odd' cases or ones which require transposition.

Plenary
Provide students with a worked example which contains errors and ask them to correct them. Ask volunteers to provide feedback for the whole class. Further examples can be given as appropriate. Alternatively, students can be asked to solve a variety of further examples and write the answers to these in their exercise books.

Lessons 23 and 24 – Solving quadratic equations by completing the square

Textbook pages 92–93

Objectives
E2.5: Derive and solve quadratic equations by completing the square.

Starter
Invite students to expand the brackets of a number of perfect algebraic squares such as $(x + 2)^2$ and $(x - 3)^2$.

Lesson commentary

- Solving quadratic equations by completing the square is not a common technique but it leads naturally on to using the formula.

- Provide students with a number of equivalent expressions such as $x^2 + 4x - 6$ and $(x + 2)^2 - 10$. Ask them to show the equivalence through expanding the perfect square and simplifying. Explain that completing the square is the reverse process of this expansion and provide a modelled example for students to follow.

- The use of a step-by-step list of instructions often helps students become confident at applying this technique since it requires a standard approach for each example. Students could be provided with further examples and asked to follow the method before writing their own list of instructions for another student to follow.

- Further consolidation and practice can be provided as appropriate. The provision of a second lesson on this topic will ensure that the students have sufficient time to complete this practice on solving equations by competing the square.

Exercise commentary

Questions 1 to 10 in exercise 32 require students to complete the square. From **question 7 onwards,** $a \neq 1$ so initial steps will be required to form an expression suitable for completing the square. These could be used as extension questions for more able students.

Question 11 uses the technique to solve quadratic equations.

Plenary
Students could be provided with three sets of cards (one with quadratic equations, one with completed square forms and one with solutions) and invited to match one card in each set to form groups of three cards. Alternatively, further examples can be given to the students who work the answers out in their exercise book before feeding back the answers to the class through discussion.

Lessons 25 and 26 – *Solving quadratic equations using the formula*

Textbook pages 90–91

Objectives
E2.5: Derive and solve quadratic equations by use of the formula.

Starter
Ask students to write down some square roots of square numbers. This will test instant recall of square number facts. Then allow the use of a calculator to find the square roots of non-square numbers. Ask students to write down these numbers to three decimal places.

Lesson commentary
- Depending on the ability of the group, the quadratic formula could be derived (following on from completing the square in the previous lesson) or simply given to the students to use immediately.

- If the formula is to be derived, students could be provided with a list of steps and asked to follow them or they could be provided with the derivation and asked to work through it. This will encourage them to think about the process more carefully.

- When it comes to *using* the formula to solve quadratic equations, explain the importance of following the rules of BIDMAS and encourage students to work 'step-by-step' rather than trying to do the calculation in one go. Ensure there is no premature rounding (work in exact form) and then encourage the final answers to be given to a sensible degree of accuracy. Three significant figures is sensible, unless otherwise quoted in the question (usually one decimal place).

- The second lesson can be used to ensure sufficient time to practise solving quadratic equations using the formula. Depending on the pace of the group and the time, further examples could be demonstrated which require transposition before a solvable quadratic is arrived at. Students could be invited to 'have a go' at solving equations of this type without too much guidance.

Exercise commentary
Exercise 30 provides plenty of routine practice in using the quadratic formula. The later examples involving fractional or decimal coefficients could be used for extension questions.

The equations in exercise 31 need to be rearranged before solving the resulting quadratic. Depending on the ability of the group, these could be used sparingly or given to more able students to provide challenge.

Plenary
Provide further examples and invite students to 'race against the clock' to write down the correct solutions. Points could be awarded for the first five correct solutions to each example.

Lesson 27 – Problems solved by quadratic equations

Textbook pages 93–96

Objectives
E2.5: Derive and solve quadratic equations.

Starter
Work through the examples on pages 93 and 94 to demonstrate how solving a problem can sometimes mean solving a quadratic equation. Focus on the algebra in example 2, as some students may find this difficult. Also discuss why one solution is sometimes meaningless and must be discarded.

Lesson commentary

- When looking at questions 1 and 2 of exercise 33, discuss with the class why it doesn't matter which of the two numbers we call x. Perhaps get one half of the class to use x and $x + 3$ in question 1, while the other half uses x and $x − 3$. Compare their results.

- Use questions 4 to 6 as another opportunity to discuss the idea of a meaningless negative solution. The class may also need reminding about Pythagoras' theorem.

- More able students might like to briefly consider what assumptions are being made in question 7, and whether it depends on where in the world Sang Jae is walking.

- There are several other areas of mathematics that students will need to remember in order to complete all the questions. These include Pythagoras' theorem, reciprocals, speed and bearings. It might be a good idea to have page references ready, so that students can look at these topics if they need to.

Exercise commentary

The examples and questions in exercise 33 provide a range of contexts in which quadratic equations need to be formed and solved. **Questions 1 to 2** deal purely with numbers. **Questions 3 to 6** deal with rectangles, and involve discarding the meaningless negative solution.

For **Question 7** students would benefit from drawing a diagram.

Questions 8 and 9 concern money. From **question 10 onwards**, students need to draw on a wider knowledge of mathematics to construct their equations. **Questions 10 and 11** involve reciprocals. **Questions 13 to 15** are about calculating speed.

Questions 16 to 20 are of a more problem-solving nature, and are significantly more challenging.

Plenary
In 2015, a GCSE maths question made the national news. As a class, search for 'Hannah's sweets' on the internet and look at the question. It is a simple probability question that results in a quadratic equation that has to be solved.

Lesson 28 – *Non-linear simultaneous equations*

Textbook pages 96–97

Objectives
E2.5: Derive and solve simultaneous equations, involving one linear and one quadratic.

Starter
Revise solving linear simultaneous equations using question 1 from exercise 19 on page 78. Together, solve this question using elimination, then solve it again using substitution. Discuss with the class which method they prefer and why.

Lesson commentary

- It is important that students understand that, while the elimination method for linear simultaneous equations is often the quickest method, often it will not work when one equation is linear and the other is not. Use example 1 on page 96 to demonstrate why it is not generally possible to eliminate a variable that is present in more than one power. Ask students to explain why this is the case.

- Using questions 1 to 3 from exercise 34 as examples, tell students that if both equations are of the form $y = \ldots$, with an expression in terms of x on the other side, the first thing they should do in their solution is equate the two right-hand sides.

- Encourage the use of substitution throughout this exercise. If students want to use a form of elimination for some of the questions, discuss this as a class and make them aware of the dangers and limitations involved in doing this.

Exercise commentary

Questions 1 to 3 in exercise 34 consist of two equations with only y on the left-hand side.

The remaining questions are not of this form.

Plenary
Give students an exam-style question based on this topic.

The revision and examination-style exercises can be used for further practice as appropriate.

CHAPTER 3
MENSURATION

Lessons 1 and 2 – Area

Textbook pages 104–108

Expected prior knowledge At this level, students should be familiar with the formulae for the area of simple geometrical figures. These lessons provide an excellent opportunity to revise these formulae and practice both problem-solving questions and ones where they have to work backwards.

The support worksheet for Chapter 3 may be useful to support lessons 1, 2 and 7 to 10.

The challenge worksheet for Chapter 3 may be useful to support lessons 5 and 6.

Objectives
E5.2: Carry out calculations involving the area of a rectangle, triangle, parallelogram and trapezium.

Starter
Ask students to write down the names of as many quadrilaterals as they can in 60 seconds. Collect answers as a group.

Lesson commentary
- A question and answer session will test students' recall of the formulae for the area of a rectangle and a triangle. Similarly, they should also be able to work out the areas of simple parallelograms and trapeziums. Trapeziums may need further revision since the formula is less familiar. Encourage students to work together and check that students are happy when the orientation of the shape is 'non-standard'.

- Introduce 'reverse' examples where missing dimensions need to be found and invite students to discuss the methods they might use. Rather than demonstrating a standard approach, encourage students to develop their own methods, the key objective being to get the correct answer.

- The second lesson will provide students with an opportunity to further consolidate this work and more able students will benefit from working on challenging questions.

Exercise commentary
Questions 1 to 4 and **questions 6 and 7** in exercise 1 are basic examples whereas **questions 5, 8 and 9** are more involved. More able students will naturally need less routine practice. **Questions 10 to 14** test the students' ability to work backwards while **questions 15 to 20** test problem-solving. These could be used as extension questions.

Exercise 2 looks at triangles and parallelograms. **Questions 1 to 15** are routine practice (some diagrams are given) while many of the other questions require an accurate sketch and the use of trigonometry. Depending on the ability of the students and their trigonometric knowledge, these could be used sparingly. The exercise could be supplemented to provide further (simple) examples for less able students. From **question 23 onwards**, students are problem-solving and working backwards to find missing dimensions. These questions could be used as extension questions for more able students or simply omitted.

Plenary
Ask students to create a compound shape and mark sufficient lengths to enable another student to work out the area of it. Students could then exchange shapes and solve each other's problems before discussing the answers at the end. Check students have correctly solved their problems and discuss any issues (not enough information, etc.).

Lessons 3 and 4 – *The circle*

Textbook pages 108–112

Expected prior knowledge Students should have met the formulae for the area and circumference of a circle before and therefore these lessons provide an ideal opportunity to revise these and work on more difficult questions.

Objectives
E5.3: Carry out calculations involving the circumference and area of a circle.

E5.5: Carry out calculations involving the areas of compound shapes.

Starter
Ask students to write down as many parts of the circle (radius, circumference, sector, diameter, arc, etc.) that they can in one minute. They can then compare answers and you can have a short discussion as to what they have written down (making sure students can define them).

Exercise commentary

Questions 1 and 2 in exercise 3 are routine practice which may need to be supplemented, depending on the ability of the group. From **question 3 onwards**, students are working with semicircles, quadrants and compound shapes. Ensure that they are setting out their working clearly in order to communicate their solutions properly.

Questions 1 to 4 in exercise 4 are examples of working backwards while **questions 5 and 6** require students to go from area to circumference or vice versa. From **question 7 onwards**, the questions test the students' ability to solve problems involving the circumference and area of circles and could be used more extensively for able groups and for extension work.

Lesson commentary

* Students could be given the diameter or radius of a circle, and then be asked to quickly work out the circumference or area using the standard formulae and write down the answers. Avoid using radius for circumference and diameter for area at this stage.

* Develop the use of the formulae for cases where diameter is given for area and radius for circumference to ensure that students are happy to work with either. This is often a common area for mistakes since students just without thinking 'plug in' the number given. Work also with a series of examples which go backwards. You could do this as a thought experiment. Ask questions like 'what would happen if you were given the circumference and asked to find the diameter?' Try to avoid any more formulae here ($d = \frac{C}{\pi}$ for example) and get students to understand the *process*.

* At this point, further practice could be given to the students during a consolidation phase of the lesson.

* Introduce more 'what if ...' situations. These could be given out on prepared worksheets and could include area of semicircle, perimeter of quadrant, etc. Students should be asked to justify their methods.

* The second lesson provides a good opportunity for further consolidation of this work.

Plenary
A quiz which not only tests going forwards with the formulae but also includes examples which need to be worked backwards. Include a semicircle or quadrant. Students could be given 30 seconds to work out the answers on their calculators for each one.

Lessons 5 and 6 – *Arc length and sector area*

Textbook pages 112–115

Objectives

E5.3: Solve problems involving the arc length and sector area as fractions of the circumference and area of a circle.

Starter

Ask students to write down the simplified fraction equivalent to $\dfrac{\theta}{360}$ for given values of θ (30, 60, 90, 120 for example).

Lesson commentary

- Begin with a series of examples where students have to work out the area and/or arc length for semicircles and quadrants. Discuss the methods that they use (divide circle area/circumference by two or four) and by referring back to the starter activity, discuss the link to $\dfrac{180}{360}$ and $\dfrac{90}{360}$.

- Ask students how they might go about finding the area/arc length for a 60° sector (divide by 6) and use this and other examples to generate a general approach to problems of this type. Formulae can be introduced at this point if necessary but encourage students to *think* about what they are doing.

- Discuss the link between arc length and the perimeter of a sector (add two radii).

- Introduce the idea of working backwards to find either the radius of the sector or the angle subtended. Ask students to write down the approach they could take in this case and encourage class discussion of the methods used. Try to avoid too much demonstration and structure unless the group require this to access the solutions.

- The second lesson provides students with an opportunity for further consolidation, additional support and the option of extension work for more able students.

Exercise commentary

Question 1 in exercise 5 is routine practice. **Questions 2 and 3** combine several sectors but should still be treated as routine practice for most students. **Question 5** develops the idea of working backwards whereas the rest of the questions test students' ability to solve problems involving arc length and sector area. These could be used sparingly for weaker groups or form the basis of the exercise for stronger students.

The next section (3.4) on chords of circles requires the use of trigonometry. This could be omitted for now since most groups will not consolidate trigonometry until Chapter 6. Alternatively, it could be used as a stretch and challenge activity or homework for more able students.

Plenary

Students could be provided with three sets of cards (one with diagrams, one with arc lengths and one with sector areas) and asked to match one card in each set to form groups of three cards. Alternatively, further questions could be given and students asked to write the answers in their exercise books for checking at the end.

Lessons 7 and 8 – Volume 1

Textbook pages 118–120

Expected prior knowledge At this level, students should be familiar with the formulae for the volume of simple solids including the cuboid and prism. These lessons provide an opportunity to revise these and attempt more difficult types of question.

Objectives
E5.4: Carry out calculations involving the volume of a cuboid, prism and cylinder.

Starter
Give students multiplication questions which include calculations where at least one of the numbers is to one decimal place. Include three numbers.

Lesson commentary

- Revise the idea of a prism and discuss the fact that cuboids and cylinders are special cases of these. Physical examples could be used from a typical household (cans, cereal boxes, chocolate boxes, etc.). Recall that the area of the cross-section can be worked out using an appropriate area formula (see previous lessons). Discuss the formula 'area of cross-section × length' for the volume of a prism and link this to the special formulae for cuboids and cylinders. Students can then practise working out the volumes of a variety of prisms using questions from the exercise. Encourage clear layout of solutions.

- Discuss the approach students might take to work out a missing dimension when presented with the volume of a cuboid or cylinder. Encourage them to develop their own approach which uses *understanding* rather than just giving alternative formulae. A collaborative approach where students are encouraged to discuss their methods with others will help deepen their understanding.

- As an extension, students could be invited to work out the surface area for some of the prisms in the exercise (**question 1 part a to d**, for example).

- The second lesson will provide students with the opportunity for further consolidation and practice.

Plenary
A calculator quiz. Give students a series of standard prisms, including cuboids and cylinders, and ask them to write down the volumes quickly in their exercise books or on mini-whiteboards if available..

Exercise commentary

Questions 1 and 2 in exercise 7 are routine practice while **questions 3 to 5** work the basic formulae backwards. The remaining questions are more of a problem-solving type and some students may struggle to determine the information required. Weaker students could be provided with further routine practice questions while more able students could be given the problem-solving questions as their main focus.

Lessons 9 and 10 – *Volume 2*

Textbook pages 120–124

Objectives
E5.4: Carry out calculations involving the volume of a sphere, pyramid and cone.

E5.5: Carry out calculations involving the volumes of compound shapes.

Starter
Fractions of ...: Ask students to find one third (for example) of a series of numbers. These could be simple numbers such as 33 or 57 or they could be decimals which divide exactly. Further examples which require the use of a calculator could also be used.

Lesson commentary

- A geometric approach to the volume of a pyramid (outlined in the textbook) can be used as an alternative to demonstration and provide students with an interesting activity. The formula for the volume of a pyramid can be derived and then compared to that of the cone ('base area' $= \pi r^2$). Students could practise applying these formulae before proceeding to the sphere. Encourage students to work together when working through problems.

- http://www.qc.edu.hk/math/Junior%20Secondary/Sphere.htm (link correct at time of publication) gives a geometric/algebraic approach to finding the formula for the volume of a sphere which may be accessible for more able students. Alternatively, a demonstration can be used or the formula simply given to the students. They should then practise applying the formula.

- Examples where the students work backwards can then be discussed. Encourage students to develop their own strategies rather than just giving alternative formulae as this will ensure understanding of the process rather than just recall.

- The second lesson provides students with more time to consolidate this work, get additional support and/or be provided with extension work.

Exercise commentary

Questions 1 to 12 in exercise 8 are routine practice while **question 13** is a typical 'ice cream cone' example. **Questions 15 to 17** require students to work backwards while the rest of the exercise contains problems to solve. Most students will be content with routine practice while more able students should be directed to the problem-solving questions from **question 18 onwards**.

Plenary
Provide students with an exam question and an incorrect solution. Ask them to work through the problem and identify the mistakes, correcting them as they go along. Pairs or groups could be given different questions and volunteers could provide feed back to the whole class.

Lessons 11 and 12 – *Surface area*

Textbook pages 125–128

Objectives
E5.1: Use current units of area, volume and capacity in practical situations and express quantities in terms of larger or smaller units.

E5.4: Carry out calculations involving the surface area of a cuboid, prism and cylinder. Carry out calculations involving the surface area of a sphere, pyramid and cone.

E5.5: Carry out calculations involving the areas and volumes of compound shapes.

Starter
Ask students to work out the surface area of a given cuboid ($3 \times 4 \times 5$ for example).

Lesson commentary

- The conceptual understanding of the formula for the surface area of a cylinder is often lacking but most students are able to apply the formula. A nice activity to try and encourage the conceptual understanding is to use a food can with the label on and ask students how they might work out the area of the label.

- Develop the formula and then allow students time to practise applying it. Encourage the students to work together if appropriate.

- When it comes to the surface area of a cone, sectors of paper can be used to demonstrate the formula or it can just be given to the students, depending on ability. Similarly, the formula for the surface area of a sphere can just be provided for the students and they can then practise using it. Encourage discussion of the methods and clear setting out of working. Reverse problems could also be introduced as appropriate for stretch and challenge.

- The second lesson provides further opportunity for consolidation and support for weaker students. More able students can be challenged to attempt harder examples.

- The short exercise which follows this work on surface area focuses on the conversion of units of area and volume. Emphasise the need for students to have a clear understanding of why the conversions are not the same as the linear equivalents but also to learn the conversions by heart.

Exercise commentary

Question 1 in exercise 9 provides basic practice in applying the formulae for surface areas.

Questions 2 to 5 are examples of reversing the process to find missing dimensions whereas **questions 6 onwards** develop a problem-solving approach. These questions could be used sparingly for less able students or form the basis of the exercise for stronger students.

Questions 1 to 15 in exercise 10 are basic practice in converting areas and volumes. **Questions 16 to 20** put the conversions in context and follow on from the work in this chapter on finding volumes and surface areas.

Plenary
As in the previous lesson, provide students with an exam question and an incorrect solution. Ask them to work through the problem and identify the mistakes, correcting them as they work. Pairs or groups could be given different questions and volunteers could feed back to the whole class.

The revision and examination-style exercises can be used for further practice as appropriate.

CHAPTER 4
GEOMETRY

Lessons 1 and 2 – Revision of standard results

Textbook pages 137–141

Expected prior knowledge Students should be familiar with basic angle facts (see starter activity) and therefore these lessons should be used to consolidate and develop this work further.

Objectives
E4.1: Use and interpret vocabulary of triangles, quadrilaterals and polygons.

E4.7: Calculate unknown angles using the following geometrical properties: angles at a point; angles at a point on a straight line and intersecting straight lines; angles formed within parallel lines; angle properties of triangles and quadrilaterals; angle properties of regular polygons; angle properties of irregular polygons.

Starter
Ask students to write down all the angle facts that they can think of. Ask them to provide an example of each one. They could write them in their books, or on poster paper if available.

Lesson commentary
• By discussion of the starter activity, ensure that the whole class have a complete set of angle facts. They should know 'angles at a point', 'angles on a straight line', 'angles in a triangle', 'X', 'F', 'Z' and 'C' angles for parallel lines and 'angles in a quadrilateral'.

• Students could then devise their own questions (and answers) which test the recall and application of the facts. Groups of students could be directed to work on different facts and then exchange their questions with students from another group.

• The angle results for polygons should also be prior knowledge but students are likely to require revision of this. A useful activity for regular polygons is to ask for a volunteer and invite them to walk the edges of a number of polygons. Ask the class to observe closely and work out the number of times the volunteer rotates their body during each walk (once only for each one). Use this demonstration to deduce the rule that the sum of the exterior angles of a polygon is always 360°.

The support worksheet for Chapter 4 may be useful to support lessons 3, 4 and 7 to 9.

The challenge worksheet for Chapter 4 may be useful to support lesson 3.

Exercise commentary
Exercise 1 tests basic recall of simple angle facts, often in an algebraic context. Stronger students should be encouraged to work through the later stages of this exercise rather than concentrating on basic practice. **Questions 17 onwards** provide a good level of challenge and **questions 23 to 26** require the sketching of a diagram.

Polygon facts are tested in exercise 2 and there are a range of questions on both irregular and regular polygons. **Question 3** is essentially the derivation of the general rule (see lesson commentary).

Exercise 3 provides a number of routine practice examples for angles derived from parallel lines. **Questions 6 to 9** provide a good level of challenge for more able students.

- Discuss the relationship between the exterior and interior angles of a polygon and use this to lead onto general rules for regular polygons. Encourage students to try to formulate their own versions of the rules before providing them with all the information. They could work on the rules in pairs or small groups if appropriate and discuss their ideas.

- Discuss the link to irregular polygons (exterior angles still sum to 360° but individual angles cannot be worked out directly). Demonstrate the triangle method and deduce the general result: sum of interior angles = $(n - 2) \times 180°$.

- Since there are lots of questions in the exercises for this topic, the second lesson can be used for additional consolidation and the provision of challenging questions for more able students.

Plenary

Students could take part in a quiz. Responses could be written on mini-whiteboards or in their exercise books. Alternatively, the questions could be true/false or multiple choice and utilise response cards.

Lessons 3 and 4 – *Pythagoras' theorem*

Textbook pages 141–143

Objectives
E6.2: Apply Pythagoras' theorem to the calculation of a side of a right-angled triangle.

Starter
Give the students a series of numbers from 1 to 12 and ask them to quickly write down the square of each number.

Lesson commentary

- Students should be familiar with Pythagoras' theorem already but if it is new material, then it could be introduced using a short investigation. Ask them to draw a right-angled triangle on squared paper and then carefully draw a square joining on to each side. Assuming the triangle is drawn in what we might call the 'standard' orientation, the two shorter sides should prove no problem and the square on the hypotenuse can be constructed using careful measurement and a protractor for the right angles. Ask them to work out the areas of the three squares and conclude that the sum of the areas of the two smaller squares is indeed equal to that of the larger square. If students have drawn a range of different sizes of triangle, the rule can be quite easily established. Dynamic geometry can also be used for this investigation if ICT facilities are available.

- Students can then practise using the theorem on a series of questions from the textbook.

- Discuss how to work out one of the shorter sides if given the length of the hypotenuse. Allow students to work out that it is subtract rather than add and then give them time to practise this application further.

- Practical applications of Pythagoras' theorem can be introduced at this point.

- The second lesson provides an opportunity for further consolidation, additional support for weaker students or the development of practical applications.

Exercise commentary
Questions 1 to 4 in exercise 4 are routine applications of the theorem, finding both the hypotenuse and a shorter side. **Questions 5 to 7** are multi-stage problems while **questions 8 to 10** require students to work algebraically. Practical applications of Pythagoras' theorem are then provided from **question 11 onwards**. Stronger students could be directed to a selection of questions from **question 8 onwards** rather than completing basic practice.

Plenary
Provide students with further examples of triangles and ask them to work out the missing sides (calculators will be useful here). This could take the form of a short test or a quiz using mini-whiteboards (if available) to display the answers. Alternatively, students can write the answers in their exercise books.

Lessons 5 and 6 – *Symmetry*

Textbook pages 144–147

Expected prior knowledge At this level, students should be familiar with the ideas of symmetry.

Objectives

E4.6: Recognise rotational and line symmetry (including order of rotational symmetry) in two dimensions. Recognise symmetry properties of the prism (including cylinder) and the pyramid (including cone).

Starter

Shape recognition: Show a series of shapes on an interactive whiteboard or overhead projector (if available) and students respond quickly with the name of the shape. Shapes could include both geometrical ones (pentagon, hexagon, rhombus, etc.) and others such as 'star', 'oval' and 'squiggle'.

Lesson commentary

- Give students a reminder of what it means to call something 'symmetrical'. Discuss the two types of symmetry and distinguish them using examples of shapes from the starter activity. Students could then try to classify the (capital, sans serif) letters of the alphabet into those which have reflective/rotational symmetry (or both). This is an ideal opportunity to discuss problems with, say, the letter 'Q' which depends how it is drawn as to whether it has a line of symmetry or not.

- Discuss the symmetrical properties of the family of quadrilaterals.

- Students should also be able to complete shapes given symmetry properties. They could be given squared paper and asked to complete a symmetrical design of their own.

- Discuss the concept of a 'plane of symmetry' (demonstration solids could be used here) and ask students to write down the number of planes of symmetry for a range of standard geometrical objects.

- The second lesson can be used for further consolidation and support for weaker students or for the provision of stretch and challenge questions for the more able.

Exercise commentary

Question 1 in exercise 5 requires students to write down the number of lines of symmetry and the order of rotational symmetry for a range of diagrams. This could be done as a class question and answer exercise. **Questions 5 to 15** test the students' understanding of the symmetrical properties of quadrilaterals and they will benefit from drawing a (neat) diagram before proceeding. These questions can be conceptually quite difficult so some pre-drawn diagrams may help less able students to make a start.

Exercise 6 looks at planes of symmetry and students will need to visualise the three-dimensional objects described in **questions 3 to 6**. If students are struggling with this, physical examples could be provided (see lesson commentary).

Plenary

Repeat the starter activity but this time ask students to write down the number of lines of symmetry and the order of rotational symmetry for the shapes provided.

Lesson 7 – Similarity

Textbook pages 147–149

Objectives

E4.1: Use and interpret geometrical terms, including similarity and congruence.

E4.4: Calculate lengths of similar figures.

Starter

Give the students a series of questions looking at multiplying two- or three-digit numbers by 2, 3, 4, etc. Include some simple decimal numbers.

Lesson commentary

- Define the term 'similar' and explain that if two shapes are mathematically similar, then one is just an enlargement of the other. Corresponding angles are equal in similar shapes but the sides are each enlarged by the same scale factor. Show the students two shapes with some sides and angles marked and ask them to work out, using these two rules, that the two shapes are similar.

- Explain that missing sides and/or angles can be found if we know that the shapes are similar.

- Provide students with some examples of similar shapes and ask them to discuss how they might fill in the missing numbers. Guide the discussion as necessary using language such as 'ratio' and 'scale factor'.

- Students can then work on consolidating this learning using questions from the exercise or other suitable examples.

Exercise commentary

Exercise 7 tests the students' ability to work with similar shapes. **Questions 1 to 4** are basic examples while **questions 5 to 11** are 'nested' triangles or other more complicated constructions. For these examples, students should be advised to draw both of the triangles separately before proceeding.

Question 12 onwards look at a more deductive use of similarity. The implications arising from these questions could be discussed as a class if appropriate.

Plenary

Give the students three further triangles with their lengths marked and ask the students to identify whether any of them are similar to each other. Ideally, two should be but the other one should not (but make it *close* in appearance to discourage direct observation).

Lessons 8 and 9 – *Area and volume of similar shapes*

Textbook pages 150–156

Objectives

E4.4: Use the relationships between areas of similar triangles, with corresponding results for similar figures and extension to volumes and surface areas of similar solids.

Starter

Ask students to multiply given numbers by a square or a cube. For example, 'What is 4×2^3?' Responses could be via mini-whiteboards, if available, or written into exercise books.

Lesson commentary

- Ask students to draw two rectangles, one 3×5 and the other 6×10. Ask them to work out the corresponding areas of the rectangles and use this to deduce the rule for the ratio of areas in similar shapes (if the scale factor of enlargement is r, the scale factor for area is r^2). Encourage students to *see* this relationship for themselves rather than just learning the rule.

- Discuss how this concept can be extended to the ratio of volumes in similar solids and provide further examples as necessary. Students should also be able to work backwards to the ratio of lengths when given corresponding areas or volumes, and also work out unknown surface areas when given the ratio of volumes. Further examples which demonstrate this can be provided as appropriate, and class discussion can be used to produce generalised approaches. Again, encourage students to *see* the relationships rather than just learning the rules.

- It is important that students work on both area and volume scale factors together to ensure appropriate linking of ideas. The first lesson may be mainly explanation and examples while the second lesson can be used to provide an opportunity for consolidation and additional support/extension questions.

Exercise commentary

Questions 1 to 6 in exercise 8 require students to work out missing areas given a ratio of side lengths while **questions 7 to 12** require them to work out missing sides when given the areas. These should be routine practice for most students. **Questions 13 to 16** introduce surface area calculations for solid shapes and the remaining questions are more involved and/or require more thought. These questions could form the basis of extension work for more able students.

Exercise 9 moves students onto volume and **questions 1 to 8** look at finding missing volumes when given side lengths. **Questions 9 to 14** require students to work out missing lengths when given volumes while **questions 15 onwards** are more interpretive. No diagrams are given and students must comprehend what the question is asking before applying the given rules. These could also be used to form the basis of extension work for more able students.

Plenary

Provide students with a range of examination-style questions and ask them to construct a model solution. Encourage them to describe the steps clearly before asking for feedback from selected students.

Lesson 10 – Congruence

Textbook page 156–157

Objectives

E4.1: Use and interpret geometrical terms, including similarity and congruence.

E4.5: Use the basic congruence criteria for triangles (SSS, ASA, SAS, RHS).

Starter

Ask the class some questions such as 'Are all rectangles similar?' (No.) and 'Are all circles similar?' (Yes.)

This can then be used to link the ideas of similarity to that of congruence in the main body of the lesson.

Lesson commentary

- Define the term 'congruent' and discuss the fact that all pairs of corresponding side lengths and angles should be equal. Discuss the conditions for demonstrating congruence in triangles and also the concept of *proof* when working with geometrical properties. Students may benefit from an example solution to a question of this type. Students working together, discussing the conditions and arguments involved, etc. is also a good way of generating a suitable approach to questions of this type.

- Formalise the ideas discussed for demonstrating congruence and provide further examples and/or questions for the students to practice.

Exercise commentary

In exercise 10, **question 1** is a simple identification exercise while **questions 2 to 8** require students to construct geometrical proofs. Encourage the drawing of a (neat) diagram before proceeding and emphasise that they must justify each assertion.

Plenary

Students who are happy with the notion of proof in this context could be provided with some examples of proofs where there is an incorrect step. They could work in pairs if appropriate, to identify the error before feeding back to the whole class or other groups.

Lessons 11, 12 and 13 – *Circle theorems*

Textbook pages 158–165

Objectives
E4.1: Use and interpret vocabulary of circles.

E4.6: Calculate unknown angles using circle theorems.

Starter
Ask students to write down the definitions of various words associated with circles (chord, radius, diameter, sector, etc.).

Lesson commentary

- Dynamic geometry software is very useful here as a demonstration tool. If ICT facilities are available, students could also be invited to try to work out some (or all) of the results themselves.

- The easiest approach is to provide the students with examples of each of the rules and ask them to produce a diagram which explains each one. They should be encouraged to write the rule in their own words. Alternatively, students could be given a series of diagrams which illustrate each rule and then asked to deduce the correct relationships through accurate measurement. Again, they should be encouraged to write down the rules in their own words.

- The key rules are 'angle at the centre is twice the angle at the edge', 'angles on the same arc are equal', 'opposite angles in a cyclic quadrilateral add up to $180°$', 'the angle in a semicircle is $90°$' and the two tangent theorems together with the alternate segment theorem. Students should be familiar with the application of all of these rules, even in quite complex diagrams. Encourage them to break the problems down and 'work around' the diagrams. Example solutions can be provided and more or less routine practice can be given.

- It is advised that students work through the circle theorems in the order suggested by the textbook. This naturally breaks down into three discrete lessons, matching the three exercises. Alternatively, the first lesson could be used to demonstrate all of the circle theorems before allowing two further lessons for sufficient consolidation and practice. Assessment points (mini-plenaries) should be spread throughout the three lessons to check student progress. These can take the form of simple question and answer sessions.

Exercise commentary

Exercise 11 provides plenty of practice in applying the first two circle theorems but from **question 10 onwards**, the diagrams are either non-standard or more involved. Students should also be familiar with isosceles triangles formed by two radii when solving these problems.

Exercise 12 provides practice of the next two rules but knowledge of the first two is assumed in several examples. From **question 13 onwards**, the diagrams are again non-standard or more involved. Depending on the ability of the class, these could be used more or less extensively.

The tangent theorems are tested in exercise 13. Again, the questions increase in complexity and students should be directed to complete examples appropriate to their ability.

Exercise 14 tests students on the use of the alternate segment theorem.

Plenary
Students could be provided with a solution to an examination-style question which contains a mistake. They should be able to work through the solution and identify the mistake, correct it and complete the solution correctly. Different questions could be issued for different students.

Lesson 14 – Constructions

Textbook pages 166–167

Objectives
E4.2: Measure and draw lines and angles. Construct a triangle given the three sides using a ruler and a pair of compasses only.

Starter
Describe a compound shape to students and ask them to recreate the shape. For example, say: 'Draw a square then add an isosceles triangle to the bottom edge of the square. From the vertex of the triangle, draw a horizontal line going to the right and then turning through a right angle to go upwards. Draw a circle inside the square.'

Lesson commentary

- Students should be familiar with the concept of constructing a triangle given the three sides and could be asked to write down a list of instructions in their own words and provide an example for another student to follow.

- Introduce the idea of geometry as a game, aiming to construct as many geometrical figures as possible, using only a pencil, a straight edge (not a ruler), and a pair of compasses. Which regular polygons can be constructed and which cannot? Which angles can be constructed and which cannot?

- Tell the students about ancient Greek mathematicians such as Euclid, Pythagoras, Archimedes and Eratosthenes.

Lesson 15 – Nets

Textbook pages 167–168

Objectives
E4.1: Use and interpret vocabulary of simple solid figures including nets.

Starter
Ask students to write down the names of as many geometrical solids as they can in two minutes. Collect a list at the end.

Lesson commentary
- Discuss different ways of representing three-dimensional objects in two dimensions. Show examples of orthographic projection, isometric projection and plan drawing (plan, front and side elevations). Comment on the need for effective methods of constructing the three-dimensional object from the two-dimensional representation. Use this to develop the idea of a net.

- Provide students with materials to construct a cube net.

Exercise commentary
Exercise 15 requires students to construct various triangles and could be used throughout the lesson to demonstrate and practise the different skills required.

Plenary
Look up how to construct a pentagon using geometry. This is quite complicated and makes a good whole-class activity.

Exercise commentary
The questions in exercise 16 are a mixture of visualisation examples and application questions. Students should be encouraged to visualise the 'folding up' of the nets and match corresponding sides to help them solve the problems.

Plenary
Provide students with examples of polyhedra such as cereal boxes, chocolate packaging, etc. Ask them to sketch the nets. Emphasise the need for the lengths to be relative to each other. They could also be asked to sketch the net of a cylinder.

The revision and examination-style exercises can be used for further practice as appropriate.

CHAPTER 5
ALGEBRA 2

Lessons 1 and 2 – *Simplifying algebraic fractions*

Textbook pages 173–175

Objectives
E2.3: Manipulate algebraic fractions. Factorise and simplify rational expressions.

Starter
Fraction cancellation: Provide students with a range of un-simplified fractions and ask them to write down the simplified equivalent.

Lesson commentary
• This looks at simplifying algebraic fractions using different techniques. Examples similar to those in the textbook can be demonstrated. Students should understand that you can only cancel if the same component appears in *all* of the terms in the fraction. Encourage them to think of terms as being separated by addition and subtraction, but not multiplication. Encourage an approach based on the factorisation of the numerator and/or the denominator before cancelling down, particularly when quadratic expressions are involved.

Lesson 3 – *Adding and subtracting algebraic fractions*

Textbook pages 175–176

Objectives
E2.3: Manipulate algebraic fractions.

Starter
Ask students to complete some simple numerical fraction additions and subtractions, writing the answers in their exercise books.

Lesson commentary
• Following on from the starter activity, encourage students to think of the lowest common denominator as the lowest common multiple of the two expressions in the denominator. Then encourage them to multiply through the numerators as appropriate. Discuss the need to then simplify the resulting single numerator if necessary.

The support worksheet for Chapter 5 may be useful to support lessons 7 and 8.

The challenge worksheet for Chapter 5 may be useful to support lessons 9 and 10.

Exercise commentary
Exercise 1 looks at (simple) algebraic fractions and students should be encouraged to simplify many of these examples 'by eye'.

Exercise 2 introduces quadratic expressions into the numerator and denominator and here students should be advised to factorise carefully first before cancelling common factors.

Plenary
Provide students with correct and incorrect examples of algebraic fractions and their simplified forms. Ask them to say whether these examples are correct or incorrect.

Exercise commentary
Exercise 3 tests the students' ability to write sums and differences of algebraic fractions as single fractions. Ensure students concentrate on finding the lowest common denominator before proceeding to multiply through into the numerators.

Plenary
Provide the students with a number of correct and incorrect examples of algebraic fraction addition and subtraction. Ask students to say whether these examples are correct or incorrect.

Lessons 4, 5 and 6 – Rearranging formulae

Textbook pages 176–183

Objectives
E2.1: Construct and rearrange complicated formulae and equations.

Starter
Provide students with some equations of the form
$ax + b = c$ and ask them to solve them quickly.

Lesson commentary

- The key idea in this series of lessons is to allow the students to work through the various different types of problem at a pace appropriate to both the needs of the class and of the individual. An approach based on demonstration should be adopted, but more able students could be encouraged to work through examples on their own before attempting further questions.

- Demonstrate a number of examples of different formulae which need transforming, up to and including $ax + b = c$ and $a(x + b) = c$. Introduce fractions, including those with x in the denominator. Introduce squares and square roots. At this level, most students should be familiar with many of these types of transformation but they will need consolidation at several points. Provide these opportunities as appropriate.

- When the new subject starts to appear twice in a formula, students often benefit from a standard method such as 'expand (or multiply), rearrange, factorise, divide' in order to put together the correct steps in the correct order. Examples of this type can be demonstrated and students should recognise the need for each step. It is often useful to ask them to write down the steps in their own words to encourage students to remember the technique.

- Ensure students are regularly checking their solutions and incorporate appropriate assessment points (mini-plenaries) throughout the series of lessons to ensure good progress is being made.

Exercise commentary

Exercise 4 provides plenty of routine practice of basic examples and can be used sparingly for stronger groups. Exercise 5 introduces fractions and negative subjects while exercise 6 develops this further to include subjects in the denominator and examples of a more complex nature. At this point, students may benefit from some more extensive consolidation before proceeding.

Squares and square roots are introduced in exercise 7 while exercise 8 contains examples where the new subject appears twice.

Exercise 9 contains examples of far greater complexity to challenge even the most able students.

All of the exercises can be used more or less extensively depending on the levels of ability within the group. Appropriate differentiation should be encouraged and students could be invited to self-select the questions they wish to attempt.

Plenary
Students could be given worked examples which contain errors and asked to work through the solutions before correcting them. This could be done in pairs or small groups if appropriate and they could exchange answers for checking.

Lesson 7 – *Variation*

Textbook pages 183–186

Objectives

E2.8: Express direct and inverse proportion in algebraic terms and use this form of expression to find unknown quantities.

Starter

Provide the students with situations, for example 8 apples cost 96 cents, and ask them to work out how much one apple would cost. Include questions which require the use of a calculator.

Lesson commentary

- Discuss the idea of proportion and introduce the alternative vocabulary in terms of variation and the symbol for proportionality. Explain that relationships between variables can be expressed algebraically if we know something about the proportional relationship.

- Demonstrate an example of direct proportion using $y = kx$ and some given values of x and y. Encourage students to follow the steps required to find the value of k. As a secondary skill, students can then work out the value of y for a different value of x (or vice versa).

- Discuss the fact that many other proportional relationships exist such as $y = kx^2$ and $y = k\sqrt{x}$. Examples can be demonstrated or students can investigate in groups.

Exercise commentary

Questions 2 to 7 in exercise 10 are routine examples of direct variation and can be used as appropriate for consolidation. **Questions 8 and 9** are more complicated examples of direct variation and could be set aside for extension work. **Questions 10 onwards** are contextualised as real-life examples but students should be encouraged to pursue a common approach to all these questions.

Plenary

Provide students with a further example in which they have to deduce the proportional relationship before answering a series of follow-on questions.

Lesson 8 – *Inverse variation*

Textbook pages 187–189

Objectives

E2.8: Express direct and inverse variation in algebraic terms and use this form of expression to find unknown quantities.

Starter

Ask students to give examples of real-life situations where two variables are in inverse proportion to each other, such as speed and time.

Lesson commentary

- Inverse variation could now be introduced. Discuss a real-life example of inverse proportion (men digging holes, for example) and demonstrate the relationship $y = \dfrac{k}{x}$ in this case. By following the methods for direct proportion, encourage students to build up a solution framework for problems of this type and again emphasise that different inverse relationships could exist such as $y = \dfrac{k}{x^2}$.

Exercise commentary

Questions 2 to 11 in exercise 11 again provide routine examples and can be used as appropriate. **Questions 12 to 15** are more complicated and could be set aside for extension work. **Questions 16 to 19** are contextualised examples but students should again be encouraged to pursue a common approach.

Plenary

Provide students with an example in which they have to deduce the inverse proportional relationship before answering follow-on questions.

Lessons 9 and 10 – *Indices*

Textbook pages 189–193

Objectives

E1.7: Understand the meaning of indices (fractional, negative and zero) and use the rules of indices.

E2.4: Use and interpret positive, negative and zero indices. Use and interpret fractional indices. Use the rules of indices.

Starter

Without any explanation, ask students to work out the answers, on a calculator, to a series of questions in which the index is negative (include one or two where it is zero too). Ask them to write down the exact values displayed on their calculators. (Ensure that they put the negative index in brackets.)

Lesson commentary

- By reference to the starter activity, introduce the idea of a negative index as a fraction, 'one over' the base raised to the positive power. Ask them to convert some of their answers in the starter to fractions and check their equivalence with the calculator answer. Deduce the value of a zero index.

- Introduce the idea of combining powers through multiplication and division by demonstrating the 'long way' of writing these kinds of sums. Expect the students to begin to recognise the patterns and deduce the rules (add for multiply and subtract for divide) themselves.

- What about the 'power of a power'? Again, by showing the 'long way' of writing these calculations, students can recognise the patterns and deduce the rule (multiply powers) for themselves.

- Use the idea of $(A^x)^2 = A$ and the rule of multiplying powers to deduce that x must be $\frac{1}{2}$ before linking to the idea of a square root. Extend this to include powers of $\frac{1}{3}$ and $\frac{1}{4}$. How could students interpret the power '$\frac{2}{3}$'? Expect 'cube root squared' or equivalent. Introduce more complicated examples of expressions which involve fractional indices and then allow students time to consolidate this work. Encourage them to check their solutions as they go along.

- Students at Extended level should have met many of the basic rules of indices before and therefore much of the initial explanation could be shortened to a series of examples and consolidation exercises. This could constitute the first of the two lessons with the second one focusing on fractional indices and more complex simplifications. Provide sufficient time for the students to consolidate this work.

Exercise commentary

Exercise 12 tests basic simplifying skills involving the rules of indices while exercise 13 brings in fractional indices as numerical evaluation problems. Algebraic simplification problems involving fractional indices are introduced in exercise 14 along with solving (simple) exponential equations (**questions 51 to 65**).

Use the questions from these exercises appropriately depending on the ability of the individual students within the group and encourage self-selection of questions.

Plenary

A short test or quiz can be used to check the students' understanding of the rules of indices covered during these lessons.

Lessons 11 and 12 – *Inequalities*

Textbook pages 193–198

Expected prior knowledge Students should be familiar with solving simple linear inequalities and representing solutions on a one-dimensional number line.

Objectives
E1.6: Demonstrate familiarity with the symbols =, ≠, >, <, ≥, ≤.

E2.5: Derive and solve simple linear inequalities.

E2.6: Represent inequalities graphically.

Starter
Provide the students with questions such as 'Does $x = 4$ satisfy $3x - 2 > 5$?' (Yes.) Expect quick responses to these questions.

Lesson commentary

- Provide revision examples of simple linear inequalities and/or consolidation of these as appropriate before moving on to two-dimensional examples. Assuming that prior knowledge is as expected, this initial activity should be short and questions carefully selected from exercises 15 and 16 to provide sufficient coverage.

- Assuming that students are confident in the drawing/plotting of straight-line graphs, the idea of shading and using dotted/solid lines to represent *in*equalities should be a straightforward concept for most to understand. Graph-drawing software is ideal for demonstrating inequalities of this type and if ICT facilities are available, students could be given the opportunity to plot examples of their own. Emphasise the need to shade out the *unwanted* sections, leaving the required region un-shaded at all times. Students should be confident in both the drawing of inequalities and the interpretation of given inequalities when represented graphically.

- A pencil-and-paper approach to this topic will mean that students take time to complete examples. As a result, the second lesson can be used to provide the additional time necessary to consolidate this work fully.

Exercise commentary

Exercise 15 is basic practice in identifying inequalities and solving simple linear inequality problems. Exercise 16 requires students to display solution sets to one-dimensional problems using a number line (ensure that students are correctly filling in the dots or leaving them open as appropriate). **Questions 20 to 25** are markedly more difficult and could be used as extension questions for more able students. From **question 26 onwards**, students are problem-solving using inequalities and these questions could be used sparingly.

In exercise 17, students are first asked to identify regions and then to draw their own. The questions can all be used for routine consolidation.

Plenary
Repeat the starter activity but this time give values of x and y and ask students whether they satisfy two-dimensional inequalities such as $x + y > 5$.

Lessons 13 and 14 – *Linear programming*

Textbook pages 198–202

Objectives
E2.6: Represent inequalities graphically and use this representation to solve simple linear programming problems.

Starter
Ask students to draw, on a set of axes (first quadrant only), a number of straight lines, for example $x = 2$, $x + y = 8$ and $2x + y = 4$.

Lesson commentary

- Describe a problem to the students similar to that given in the textbook. Explain that the information can be interpreted algebraically into a number of *constraints* which can be displayed on a graph as a series of inequalities (see previous lesson). Demonstrate the solution (using graph-drawing software if appropriate) and discuss the significance of the points contained within the correct region. What does each one mean in reality? Can we interpret the values shown? How do they relate to cost/profit? What are the maximum/minimum values?

- Students can then attempt a further example, working together or in formal pairs or small groups, if appropriate. If ICT facilities are available students could also solve problems of this type using graph-drawing software to speed up the process of drawing all of the graphs by hand.

- If a hand-drawn approach is used and/or if further consolidation is needed, the second lesson can be used to provide time to practise solving further linear programming problems.

Exercise commentary
Exercise 18 contains a range of examples, the first three of which are not contextualised and can be used for basic practice. The remaining questions provide a practical context for students to first interpret and determine the issues before proceeding to draw the graphs and work out the solution. Different pairs or groups of students could be given different questions to try depending on ability.

Plenary
Provide students with a graph showing the solution to a linear programming problem and ask them to write down a real-life situation which it could represent. They will need to interpret the inequalities as constraints and then put these into words. Encourage students to work together and ask for examples from the group at the end.

The revision and examination-style exercises can be used for further practice as appropriate.

CHAPTER 6
TRIGONOMETRY

Lessons 1 and 2 – Right-angled triangles

Textbook pages 208–213

Expected prior knowledge Students should have met right-angled triangle trigonometry before so these lessons provide a good opportunity to thoroughly revise these ideas and develop them further.

Objectives
E6.2: Apply the sine, cosine and tangent ratios for acute angles to the calculation of a side or of an angle of a right-angled triangle.

Starter
Division to decimals: Make sure that students have a calculator and then give them a number of (exact) divisions which lead to a decimal answer between zero and one ($5 \div 8$, $6 \div 10$, etc.).

Lesson commentary
- Depending on the students' previous knowledge of trigonometry, most of the introductory work here could be omitted and the lessons would provide time for a series of revision examples and exercise practice.

- Show the students a diagram of a right-angled triangle with one of the acute angles marked. Deal with the terminology of trigonometry (opposite, adjacent and hypotenuse) first. Then you could proceed in one of two ways. Assuming this is the first time the students have met the idea of trigonometry, you could get them all to accurately draw a right-angled triangle with a fixed acute angle (say 30°). The idea is that they all draw different-sized triangles and you can then get them to work out the ratios $\frac{O}{A}$, $\frac{O}{H}$ and $\frac{A}{H}$. By comparing them to each other's answers, they should be (about) the same.

- This can then lead into a discussion about the ratios sine, cosine and tangent representing these three measured ratios for a given angle (change the angle, change the ratio). Students can then be introduced to the buttons on their calculators for sin, cos and tan before working out the value of say sin(30°), cos(45°) or tan(70°). Encourage them to write these down accurately (full calculator display) before rounding appropriately.

The support worksheet for Chapter 6 may be useful to support lessons 1 and 2.

The challenge worksheet for Chapter 6 may be useful to support lessons 2, 9 and 10.

Exercise commentary

Exercise 1 is a simple investigation into 'tangent' and could be used as part of the introductory activity.

In exercise 2, students are required to find the missing side when it is in the numerator of the ratio (**questions 1 to 15**) and when it is in the denominator of the ratio (**questions 16 to 22**). **Questions 23 to 34** have no diagrams provided and students should be advised to draw a neat, labelled diagram before proceeding to find the missing side length.

Exercise 3 involves multi-stage problems and could be set for extension work or as part of a further session set aside for consolidating the learning so far.

- An alternative approach is to use clinometers and long tape measures or trundle wheels and say to the class that they are going to survey certain landmarks around the school (clock tower, large tree, etc.) Ask them to record the angle of elevation together with the distance that they are away from the landmark.

- When they have completed the survey, explain that they can use the ideas of trigonometry to find the heights of their landmarks. Discuss the *tangent* ratio and explain that this is the ratio of the opposite side (their unknown height) to the adjacent side for a *given angle*. If they can work this ratio out ('tan' button on their calculator), they can use the value of the ratio to find the missing heights.

- Students can then proceed to carry out the calculations using the formula for tan.

- Explain that the sine and cosine are just two other ratios (for different combinations of sides) and that they can also be useful in solving problems involving right-angled triangles when an angle is given.

- The students should now begin to formalise their methods for finding missing side lengths using the trigonometrical ratios. Provide them with some examples and ask them to follow the steps in the working. Most students will need at least one example of each ratio, but encourage them to label the sides in order to correctly identify the appropriate ratios for themselves. Introduce them to the 'SOHCAHTOA' idea as a good way of remembering the ratios or use a triangle similar to that used for speed = distance/time.

- Provide the students with plenty of opportunity to practise working out missing sides where the unknown is in the numerator. Then move on to demonstrate the finding of a side which appears in the denominator. Suggest easy ways to solve these types of problem (swap the unknown and the trigonometrical ratio or 'denominator = divide', for example). Again, provide students with the opportunity to practise examples of this type.

Plenary

Revise the vocabulary of trigonometry by getting students to write down their own definitions of the key words encountered. They could then discuss their definitions with another student to see if they make sense.

Lessons 3 and 4 – *Find an angle*

Textbook pages 214–216

Expected prior knowledge Students should be familiar with finding angles using trigonometry so these lessons provide an ideal opportunity to consolidate this work and develop it further. The starter and lesson commentary provide a framework for re-introducing the topic.

Objectives
E6.2: Apply the sine, cosine and tangent ratios for acute angles to the calculation of a side or of an angle of a right-angled triangle.

Starter
Give the students a series of decimal values and ask them to use the 'invsin', 'invcos' or 'invtan' buttons on their calculators to write down, on mini-whiteboards or in their exercise books, the values they get (rounded to one decimal place as appropriate).

Lesson commentary

- Students may question why they have done the starter activity and what it is providing. This is a good introduction to the idea of finding the angle in a right-angled triangle when given two of the sides. Explain the purpose of the buttons they have used, and then demonstrate a few examples where the ratio of sides is worked out and then the inverse buttons used to work out the angle. Again, emphasise the appropriate labelling of the triangle and the need to ensure the correct ratio is selected.

- Students can then proceed to try examples of their own; working together should be encouraged, if appropriate.

- The second lesson provides further opportunity to consolidate all of this work on trigonometry and ensure that all students have completed sufficient consolidation. More able students can use this time to attempt more complex problems.

Exercise commentary
Exercise 4 provides plenty of basic practice in finding angles. These questions can be used as appropriate. **Questions 16 to 20** have no diagrams so students should be advised to draw a neat diagram before proceeding. **Questions 21 to 26** are multi-stage problems and could be used as extension work.

Plenary
Give the students a card-matching activity to carry out in pairs. Three sets of cards could be issued, one set with diagrams of right-angled triangles with a missing side or angle, one set with 'workings out' and one set with the answers. Ask them to correctly match the cards into threes. Alternatively, provide the students with further examples and ask them to quickly work out the answers and write them in their exercise books.

Lessons 5 and 6 – *Bearings and scale drawings*

Textbook pages 216–220

Expected prior knowledge Students should be familiar with bearings so these lessons provide an ideal opportunity to revise and consolidate these ideas.

Objectives

E4.3: Read and make scale drawings.

E6.1: Interpret and use three-figure bearings.

E6.2: Solve trigonometric problems in two dimensions involving angles of elevation and depression.

Starter

Complementary angles: Give the students a series of angles and ask them to work out the difference between these angles and 360°.

Exercise commentary

The questions in exercise 5 are not exclusively about bearings but are of a general problem-solving type. Students should be encouraged to draw a neat, labelled diagram before proceeding. The problems become more difficult and students could be directed to complete certain questions, depending on ability.

Exercise 6 on scale drawing links into this work on bearings and trigonometry and could be used as a further exercise if appropriate.

Lesson commentary

- Discuss the various methods we might use to recognise where we are and to know where we are going. We could use a map, landmarks, Satellite Navigation, a list of directions or just our own knowledge. But what if we have none of these traditional methods of navigation? This leads onto a discussion about bearings and their use by ships and aircraft (and walkers).

- Students may have some familiarity with bearings but it is important that they understand the conventions clearly. 'Clockwise from North' and three-figure bearings at all times are the two important facts that students should be able to recall.

- Provide them with a series of examples and ask them to write down the bearings that are given. Discuss the idea of 'back-bearings' and how they might be calculated relative to each other.

- Students could then be given the task of drawing their own bearings and incorporating this into the making of a scale drawing. This could be from information given or they could design their own, possibly in the form of a 'treasure hunt'.

- Link back to the work done previously on trigonometry and Pythagoras' theorem and demonstrate a practical situation where these are used to solve a problem involving bearings (the example in the textbook is a classic situation).

- Provide the students with further examples as necessary and then get them to try to solve some problems of their own. Encourage students to work together throughout, if appropriate.

- The second lesson will allow students sufficient time to complete consolidation questions, particularly when the drawing of diagrams is required.

Plenary

Give the students a compass direction and ask them to give the bearing associated with it (North is 000° and South East is 135° for example).

Lessons 7 and 8 – *Three-dimensional problems*

Textbook pages 220–222

Objectives
E6.5: Solve simple trigonometrical problems in three dimensions including angle between a line and a plane.

Starter
Pythagoras' theorem revision: Give the students a series of questions which revise the application of Pythagoras' theorem for standard right-angled triangles with the hypotenuse as the unknown. Students could work out the answers quickly on a calculator and display them on mini-whiteboards, if available, or write the answers in their exercise books.

Lesson commentary
- Students often have difficulty in visualising three-dimensional problems involving Pythagoras and trigonometry. The classroom is a good place to start trying to visualise three-dimensions. Questions such as 'what is the longest pole that could fit into the classroom?' are ideal for helping students with this visualisation since they can easily picture the situation.

- Explain that three-dimensional problems are just a series of multi-stage two-dimensional problems. Encourage students to 'extract' right-angled triangles from their three-dimensional situations. Encourage careful labelling and discourage students from simply learning the concepts without knowing how to apply them. Students should be able to use the appropriate trigonometrical ratios and/or Pythagoras' theorem when required.

- Problems can be demonstrated for the students to follow or students could be given questions and asked to discuss them in pairs or small groups. Students should be encouraged to work together and discussion of the various situations should be encouraged throughout this work, if appropriate.

- The second lesson will allow students time to fully consolidate this topic through further question practice.

Exercise commentary
Exercise 7 contains a series of commonly encountered situations involving cuboids, pyramids and 'wedges'. Ensure students are identifying the correct two-dimensional triangles when working out the answers to the problems. **Question 6** has no diagram so students should be encouraged to sketch the cuboid before carefully labelling it and then proceeding to solve the problem. From **question 8 onwards**, the problems are more applied and these could be used to extend more able students. Again, diagrams are sometimes not provided so students should sketch the situation before proceeding.

Plenary
Students could be given an examination-style question and the solution and be asked to describe the method of solution. Alternatively, incorrect solutions could be given to students, the expectation being that they can correct the errors and produce a complete, correct solution.

Lessons 9 and 10 – *Sine, cosine and tangent for any angle*

Textbook pages 222–224

Objectives

E6.3: Recognise, sketch and interpret graphs of simple trigonometric functions. Graph and know the properties of trigonometric functions. Solve simple trigonometric equations for values between 0° and 360°.

Starter

Ask students to write down the values of sine, cosine or tangent for a range of different acute angles given. Calculators should be used.

Lesson commentary

- Explain that work with right-angled triangles is an *application* of trigonometry rather than the whole purpose of it. Discuss the nature of angles greater than 90° and explain that the sine, cosine and tangent ratios exist for *any* angle.

- Ask students to create a table of values for angles between 0° and 360°, going up in 15° or 30°. Ask them to use their calculators to work out the values of sine, cosine and tangent for these angles (tan(90°) and tan(270°) are undefined and students could be asked to consider *why*). Students can then proceed to plot the graphs of the three trigonometrical ratios (on separate axes). Encourage them to look at the symmetry of the graphs (they could write this down from observations or discuss it with a partner).

- Demonstrate how the curves can be used to find out the answers to trigonometrical equations such as $\sin x = -0.5$ and then allow students to practise examples of their own. The second lesson will allow sufficient time for consolidation, particularly if the graph-drawing approach suggested here is used to develop this work.

Exercise commentary

Questions 1 to 3 in exercise 8 can be used as an introductory activity as described in the lesson commentary. **Questions 4 to 15** use the symmetries of the graphs to solve simple trigonometrical equations.

Plenary

Students can be invited to give two angles in the range 0 to 360° for a series of further trigonometrical equations. Responses could be on mini-whiteboards or they could write down the answers in their exercise books for checking at the end.

Lessons 11 and 12 – *The sine rule*

Textbook pages 225–227

Objectives
E6.4: Solve problems using the sine rule for any triangle.

Starter
Ask students to write down the value of sin*x* for a series of given angles where *x* is between 0° and 180°.

Lesson commentary

- Give the students the sine rule as a simple formula and then demonstrate various problems which use it before allowing them time to practise. If this is the preferred approach, make sure that the range of problems covers finding both angles and side lengths and it is also important to include a more complicated or applied problem.

- If the students are able, it is more satisfying to demonstrate the derivation of the sine rule (or get them to derive it themselves with some guidance and through discussing together if appropriate). The problems then become more 'real', and students can see the benefit of the sine rule over a long-hand approach using right-angled triangles. Discuss also how the sine rule reduces to the standard $S = \dfrac{O}{H}$ when one of the angles being considered is 90°.

- The second lesson can be used to ensure that students have sufficient time to consolidate this work on using the sine rule in situations where triangles are not right-angled.

Exercise commentary

Questions 1 to 6 in exercise 9 are standard routine practice examples of finding side lengths while **questions 7 to 10** require clearly labelled diagrams before proceeding. Likewise, **questions 11 to 18** are routine practice examples of finding missing angles while **questions 19 to 22** require diagrams.

Plenary
Students could be asked to fill in the blanks in a series of partially completed worked examples. They could work in pairs or small groups, if appropriate, to complete the problems before feeding back to the whole class. Alternatively, a further example could be given to the class to solve in their exercise books.

Lessons 13 and 14 – *The cosine rule*

Textbook pages 227–231

Objectives
E6.4: Solve problems using the cosine rule for any triangle.

Starter
Ask students to write down the value of cos$x$ for a series of given angles where x is between $0°$ and $180°$.

Lesson commentary
- Give the students the cosine rule as a simple formula and then demonstrate various problems which use it before allowing them time to practise. If this is the preferred approach, make sure that the range of problems covers finding both angles and side lengths and it is also important to include a more complicated or applied problem.

- If the students are able, it is better to demonstrate the derivation of the cosine rule (or get them to derive it themselves with some guidance and through working together). The problems then become more 'real', and students can see the benefit of the cosine rule over a long-hand approach using right-angled triangles. Discuss also how the cosine rule reduces to the standard $C = \dfrac{A}{H}$ when the angle being considered is $90°$ $(\cos(90°) = 0)$.

- The second lesson will ensure that students have sufficient time to consolidate this work on the cosine rule and attempt questions that are more problem-solving in nature.

Exercise commentary
Questions 1 to 6 in exercise 10 are standard routine practice examples of finding side lengths while **questions 7 to 10** require clearly labelled diagrams before proceeding. Likewise, **questions 11 to 16** are routine practice examples of finding missing angles while **questions 17 to 21** require diagrams. **Questions 22 to 24** are further examples of finding side lengths.

Exercise 11 is a mixed exercise of applied problems which use the cosine rule and students should be advised to draw a clearly labelled diagram before proceeding. Students could be invited to select the questions they wish to attempt with advice that they generally get harder as the exercise goes on.

Plenary
Students could be asked to fill in the blanks in a series of partially completed worked examples. They could work in pairs or small groups, if appropriate, to complete the problems before feeding back to the whole class. Alternatively, a further example can be given to the students for them to solve in their exercise books.

The revision and examination-style exercises can be used for further practice as appropriate.

CHAPTER 7
GRAPHS

Lessons 1 and 2 – Straight-line graphs

Textbook pages 239–246

Expected prior knowledge Students should be familiar with work on the equation of a straight-line graph.

Objectives

E3.1: Demonstrate familiarity with Cartesian coordinates in two dimensions.

E3.2: Find the gradient of a straight line. Calculate the gradient of a straight line from the coordinates of two points on it.

E3.3: Calculate the length and the coordinates of the midpoint of a straight line from the coordinates of its end points.

E3.4: Interpret and obtain the equation of a straight-line graph.

E3.5: Determine the equation of a straight line parallel to a given line.

E3.6: Find the gradient of parallel and perpendicular lines.

Starter

Provide students with several examples of the form $y = mx + c$ and get them to substitute various values of x into them.

Lesson commentary

- Distinguish between the idea of *plotting* graphs and *sketching* graphs. When plotting, it is important to make a table of values and then plot each coordinate accurately on the grid before joining up the points with a neat, straight line. A sketch should represent the line accurately (axis intersections, for example) but does not need to be formally plotted. Different lines should be correct relative to each other.

- Discuss the formal equation of a straight-line graph, $y = mx + c$, and the idea of the gradient and intercept. Students should be able to work out the gradient from a triangle based on two known coordinates, and write down the equation of the straight line from this. More practice in using one of the coordinates to find c may be useful.

- Discuss the idea behind the gradients of perpendicular lines. Students should see the relationship and be encouraged to work with the formula $m_1 \times m_2 = -1$. Explain that once the gradient of the perpendicular line has been found, the method for finding the equation is the same as before.

- A mixture of questions from the various exercises can then be used to test students' understanding of this topic. Careful selection of questions from the exercises will enable students to consolidate this work fully.

The support worksheet for Chapter 7 may be useful to support lessons 7 and 8.

The challenge worksheet for Chapter 7 may be useful to support lessons 10 and 11.

Exercise commentary

Exercise 1 concentrates on the accurate plotting of straight lines given both explicitly and implicitly. **Questions 17 to 20** are applied problems.

Exercise 2 concerns finding the gradient of line segments when given coordinates. Exercise 3 deals with $y = mx + c$ formally and students sketch the lines given. From **question 13 onwards**, lines are defined implicitly and will require rearrangement before proceeding.

Finding the equation of a line is dealt with in exercises 4 and 5.

Exercise 6 deals with finding the equation of a line perpendicular to the given line.

Plenary

Provide students with two sets of cards, one set containing graphs and the other containing a series of equations of straight lines, defined both explicitly and implicitly.

Ask students to match the cards from each set to create pairs. Alternatively, further examples similar to those in the exercises can be given to students to assess progress.

Lessons 3, 4 and 5 – Plotting curves

Textbook pages 246–253

Objectives
E2.11: Construct tables of values and draw graphs for functions of x.
E2.12: Estimate gradients of curves by drawing tangents.

Starter
Provide students with a number of formulae involving x^2 and ask them to substitute given values into them. Include negative values of x.

Lesson commentary
- Students should be familiar with the idea of plotting simple curves but may need some practice in setting out the working clearly in a table. Provide them with an example similar to that in the textbook and ask them to plot the curve given. Encourage them to work together and check each other for accuracy. Further examples, including ones that involve $\frac{a}{x}$ can then be provided as appropriate.

- Explain that, unlike straight-line graphs, the gradient of curved graphs changes constantly as you work along the curve. Ask for suggestions as to how the gradient at a particular point might be *estimated*. Deduce that a tangent drawn to the curve at the given point can give us a guide to the gradient at that point. Emphasise that it is just an estimate since we are relying on the accuracy of both the plotted graph and our tangent. Refer back to the previous lesson, if necessary, in order to work out the gradient of the resulting tangent line. Then make the students practise plotting graphs and working out gradients.

- If time permits, a brief discussion of the shapes of some standard functions could be undertaken.

- The number of questions in the various exercises associated with this topic, and the amount of time needed for accurately plotting curves, does not make this a quick topic to consolidate. The second and third lessons in the sequence can therefore be used to ensure students have enough time to complete a range of examples.

Exercise commentary
Exercise 7 focuses on plotting the graphs of quadratic functions. Note that function notation is used from **question 14 onwards** but students should be advised that this is just the same as '$y = ...$'. Encourage the formal use of a table of values throughout.

In exercise 8, **questions 1 to 3** look at the idea of finding gradients from plotted graphs while **questions 4 to 17** focus on plotting graphs of more complex functions. **Question 18** returns to the idea of using the graph to find the gradient while **question 19** requires students to plot a complex function. **Questions 18 to 20** can be used as extension questions for more able students.

Exercise 9 puts the work on exponential functions into context.

Exercise 10 puts much of this work into context by using functions relating to real-life problems. Students are invited to draw graphs of the given functions before using the graphs to solve follow-on problems. **Questions 5 and 6** deal with exponential functions.

Plenary
Ask students to illustrate, using their arms, a series of standard graphs such as $y = x^2$ and $y = x^3$. This activity is known as graph aerobics.

Lesson 6 – *Interpreting graphs*

Textbook pages 254–255

Objectives

E2.10: Interpret and use graphs in practical situations including travel graphs and conversion graphs. Draw graphs from given data.

Starter

Speed calculations: Give the students a distance and a time and ask them to work out the speed. This can be in the form of a quiz with mini-whiteboards used for the responses or the students can write down the answers in their exercise books for checking at the end.

Lesson commentary

- Discuss common 'real-life' graphs that might be encountered (distance graphs, temperature graphs, conversion graphs, share prices, etc.) then provide the students with a real-life graph (projected onto a screen, if available) and ask them questions about it. Questions could include 'how much ...' and 'how long ...', depending on the exact nature of the graph.

- Ask students to work through some of the other types of real-life graphs from exercise 11, including conversion graphs. Students could look at these in pairs or small groups if appropriate, but encourage good, clear interpretations and sound reasoning throughout.

- Consider a situation such as a car hire company that charges a standard rate of $5 and then 10 cents per kilometre thereafter. How might we draw a graph of this situation? What would be the cost of hiring a car and travelling 200 kilometres in it?

Exercise commentary

Questions 1 and 2 of exercise 11 are simple 'interpret' questions where students are reading from given graphs while in **questions 3 to 6** they are expected to draw the graphs first before using them to solve follow-on problems. Ensure that students are completing the graphs accurately before using them.

Plenary

Ask students to invent their own situation similar to those in the exercise and challenge a partner to solve problems based around it. Some of these situations could be shared with the whole class.

Lessons 7 and 8 – *Graphical solution of equations*

Textbook pages 255–260

Objectives
E2.11: Solve equations approximately, including finding and determining roots by graphical methods.

Starter
Quadratic substitution: Provide students with a quadratic expression and ask them to substitute a series of numbers into it. These could be generated by the roll of a die. If appropriate, move on to include negative numbers.

Lesson commentary

- Provide students with a prepared quadratic graph (projected onto a screen, if available) and ask them how they might use it to solve the equation equal to zero. Discuss the idea that the graph traces the values of 'y' for a range of x-values and therefore when we are looking at an equation such as $x^2 + 3x - 1 = 7$, we are really looking at when the graph has a y-value of 7: go across from $y = 7$ to the curve and read down to find the x-values which *approximately* solve the equation. These can be checked by substituting the values back into the equation.

- Ask the following questions: What could we do if the quadratic expression were equal to a linear function rather than just a number? How could we use the graph if the equation is not the same as the function in our original graph? Students should be encouraged to adopt a systematic approach to examples of this kind. Emphasise that it is important that the left-hand side is exactly the given function for which the graph is drawn (transposition may be necessary), and that the right-hand side should be the supplementary function drawn on.

- Discuss this approach for further examples, including ones which are not quadratic.

- The second lesson will enable students to have sufficient time to consolidate this work, particularly since there is a lot of graph-drawing involved. Select questions from the exercise as appropriate to ensure sufficient practice is given.

Exercise commentary
Questions **2 to 4** in exercise 12 test students' ability to draw neat graphs and use them to solve the given equations. They can be considered routine practice questions. In **questions 5 to 9**, students must switch the given function to determine the supplementary function that would need to be drawn. The rest of the exercise develops these ideas further with students drawing graphs and supplementary functions and solving equations as a result. Students could be directed to look at the later questions if particularly able but most students should concentrate on **questions 10 and 11**.

Plenary
Project a quadratic graph onto a screen, possibly using graph-drawing software, and ask students to read off the approximate solutions to a series of questions of the form 'What is x if $y = 3$?' Alternatively, prepared worksheets can be given out.

Lessons 9, 10 and 11 – Distance–time and speed–time graphs

Textbook pages 260–266

Objectives
E2.10: Apply the idea of rate of change to easy kinematics involving distance–time and speed–time graphs, acceleration and deceleration. Calculate distance travelled as area under a linear speed–time graph.

Starter
Speed/distance/time calculations: Give the students a distance and a time (or speed/time, distance/speed) and ask them to work out the speed (distance/time). This can be in the form of a quiz with mini-whiteboards used for the responses or students can write down the answers in their exercise books.

Lesson commentary

- Students should be very familiar with the concept of a distance–time graph but may not be overly familiar with the idea that the *gradient* of the graph gives the speed. The idea can be clarified and developed by comparing the formula for speed and the use of a triangle when finding the gradient (the *y*-change is the distance and the *x*-change is the time).

- Speed–time graphs will be less familiar to students but these can be developed naturally from the idea of a distance–time graph. Provide students with an example of a speed–time graph and ask them to describe what is happening. They could be encouraged to work in pairs or small groups and discuss this together, if appropriate. Then provide them with an example and ask them to draw the resulting speed–time graph.

- Discuss the idea that change in speed (over time) can be described as acceleration and discuss the similarities with speed being the change in distance over time. Clarify the idea that acceleration can be worked out by looking at the gradient of the speed–time graph at certain points.

- Suggest the question: 'If I have travelled for 10 seconds at 30 m/s, how far have I travelled?' (300 metres.) Link this to the area below the line

(a rectangle 10 by 30) on a speed–time graph representing this constant speed. Explain that, in general, the area beneath a speed–time graph will give us the total distance travelled. Include further examples where speed is not constant (areas will then include triangles as well as rectangles).

- Since this work all links together, it is useful to consider it as a single topic. As a result, several lessons are likely to be needed to ensure complete coverage. The textbook provides a natural split between distance/time and speed/time graphs so the first lesson could be used for the distance/time work and the second lesson for work on speed/time graphs. The third lesson will provide an ideal opportunity for students to fully consolidate both aspects of this work. Appropriate assessment points (mini-plenaries) can be incorporated throughout this series of lessons.

Exercise commentary
Questions 1 and 2 in exercise 13 are comprehension questions and could be completed using whole-class discussion. Students are expected to draw the graphs from the scenarios described in questions 3 to 7 while questions 8 and 9 return to comprehension but for more complicated situations. In exercise 14, students are expected to use the given graphs to work out acceleration, distance and average speed. Emphasise the need for correct use of units.

Exercise 15 contains a series of questions where students must sketch speed–time graphs from the information given. In questions 12 and 13, the acceleration is non-uniform and students must use the idea of a tangent to the curve to estimate acceleration.

Plenary
Provide the students with one further example of a speed–time or distance–time graph and ask them questions about it. Encourage them to discuss the example as a class when answering the questions provided.

Lesson 12 – *Differentiation*

Textbook pages 266–269

Objectives
E2.13: Understand the idea of a derived function.

Starter
Ask the class to tell you what they understand by the idea of a gradient. Ask them how they would answer the questions 'What is the gradient of the curve $y = x^2$?' Discuss the idea that it would depend where on the curve you were.

Lesson commentary

- Introduce the idea of calculus, differentiation and derived functions. You don't necessarily need to teach students how to differentiate from first principles at this point, as sometimes this has the effect of making them think that what is quite a simple topic is going to be far more difficult than it is. You may, however, wish to do this if it is a particularly able class.

- Students should be taught both notations for the derived function, as they will both be used. Also familiarise them with the phrase 'with respect to x' and what it means.

- It may be a good idea to work through an example containing fractions with the class, just to make sure that nobody has any problems with dealing with fractions.

- Tell students that to differentiate a function containing multiple terms, you simply differentiate each term. Do not attempt to prove why this works at this stage.

- Students should also be made aware of mental shortcuts that they will use, such as that differentiating a constant will always give zero, and differentiating nx will just give n. In these cases, they do not have to think through the process of multiplying by the power and subtracting 1 from the power.

Exercise commentary
Exercise 17 is very straightforward and just contains functions to be differentiated. Exercise 18 is slightly more complicated. **Questions 1 to 3** involve setting the derived function equal to a number and then solving it. **Questions 4 and 5** introduce more algebra. **Question 6** has a negative x^2 term, so will require extra care. The remaining questions involve higher powers of x and can be used as extension questions.

Plenary
Teach the class a little bit about the history of calculus and the controversy surrounding Newton and Leibniz.

Lesson 13 – *Turning points*

Textbook pages 266–267, 270–271

Objectives
E2.13: Apply differentiation to gradients and turning points (stationary points). Discriminate between maxima and minima by any method.

Starter
Discuss how it might be possible to find the maximum or minimum values of a function by completing the square. Tell students that unfortunately this method cannot be used on cubic equations. Show students the graph of a cubic function, define and identify a turning point, then ask students what the gradient of the curve is at this point.

Lesson commentary

- There is quite a lot of vocabulary that needs to be clearly explained at the start of this lesson: 'turning point', 'stationary point', '(local) maximum and minimum points'. Remind students what is meant by a 'cubic function'.

- Discuss the number of turning points it is possible for quadratic and cubic functions to have.

- The idea of setting the derived function to zero and solving it to find turning points should be clearly demonstrated. Students should fully understand why it is the derived function, not the original function, that needs to be set equal to zero.

- Explain why the nature of a turning point can be determined using the second derivative. Students should appreciate that although it seems backwards (i.e. a negative value of the second derivative implies a maximum point), what it really means is that at a maximum point, the gradient is decreasing.

Exercise commentary
Use exercise 16 to introduce the idea of turning points – the questions provide practice in finding the coordinates of the minimum or maximum point on the graph of a quadratic function by completing the square.

Questions 1 to 6 of exercise 19 ask students to find turning points and determine their nature. **Questions 7 and 8** are extension questions and involve the idea of a discriminant, which students may need reminding about.

Plenary
Can students think of any stationary points that are neither a maximum nor a minimum? Discuss the idea of stationary points of inflection with the class.

The revision and examination-style exercises can be used for further practice as appropriate.

CHAPTER 8
SETS, VECTORS, FUNCTIONS AND TRANSFORMATIONS

Lessons 1, 2 and 3 – *Sets*

Textbook pages 280–289

Objectives
E1.2: Use language, notation and Venn diagrams to describe sets and represent relationships between sets. Definition of sets e.g. $A \{x: x$ is a natural number$\}$, $B \{(x, y): y = mx + c\}$, $C \{x: a \leq x \leq b\}$, $D \{a, b, c, ...\}$.

Note: *since there are several aspects to this work on sets and lots of practice questions in the associated exercises, it is envisaged that this work will take three lessons to complete. The first lesson in the series is likely to be an introduction to the terminology and notation. The second lesson will enable students to develop the idea of using Venn diagrams and the third lesson will provide the opportunity to consolidate this work fully and solve problems using sets. Careful selection of questions will ensure students have sufficient practice to consolidate their knowledge. Regular assessment points (miniplenaries) are included to help assess student progress.*

Starter
Number classification: Ask the students to classify the numbers from 1 to 20 into sets such as 'prime numbers', 'even numbers' and 'multiples of four'.

Lesson commentary
- Following on from the starter activity, describe how objects can be classified according to certain characteristics and that these groups are called sets. The letters of the alphabet can be classified according to their symmetry properties, for example.

- Students in the class could be classified according to their hair colour or whether they wear glasses. Use these or similar examples to outline the various terminologies and notations associated with sets including 'intersection', 'union', 'subset', 'member/element', 'universal set' and 'complement'.

The support worksheet for Chapter 8 may be useful to support lessons 6 and 7.

The challenge worksheet for Chapter 8 may be useful to support lessons 6 and 7.

Exercise commentary

Exercise 1 asks students to interpret information given in a series of Venn diagrams and write down the number of things falling into various sets.

Exercise 2 uses more formal set notation and asks students to record the various subsets and answer true or false questions.

The use of Venn diagrams to show various regions follows in exercise 3. Students are invited to draw multiple copies of the diagrams given and shade in regions which satisfy certain conditions. Later parts of both **questions 4 and 5** are quite complicated and could be used as extension work for the more able. In **question 7**, students are expected to write down the regions shown in the diagrams.

Exercise 4 contains many examples of logical problems which can be solved using Venn diagrams and/or sets. **Questions 1 and 2** are scaffolded to guide students through the process but from **question 3 onwards** they must work out the steps of logic themselves. Students should be encouraged to work together here using a variety of approaches. Emphasise that it is the answer that is all important. **Questions 15 to 19** require students to express various logical statements using formal set notation and could be set aside for extension work.

- Introduce students to the idea of Venn diagrams and suggest how these can be used to provide a visual representation of these terminologies. Students could be asked to write their own glossaries of the terms and provide their own examples for their definitions.

- Encourage students to use the notation and terminology carefully when describing sets and provide them with a range of examples where they are using given Venn diagrams and/or sets to solve a variety of problems.

- Formalising the use of Venn diagrams to show various regions should follow next. How would we show, for example, the intersection of A and not B? Students could be encouraged at this point to discuss, in pairs or small groups, what each of the regions in both two- and three-circle Venn diagrams represent. For example, A and B but not C translates into $A \cap B \cap C'$.

- The use of Venn diagrams and the ideas behind sets can be used to solve various logical problems, examples of which are given in section 8.2 in the textbook. Demonstrate a number of examples and encourage the students to follow the logical steps in the deduction process. Ask them to describe each of the stages in their own words. Provide further examples and ask the students to work together to form a correct logical argument and solve the 'puzzle'. Encourage formal terminology and notation, but emphasise that it is more important to get the right answer than worry about absolute precision.

Plenary

Ask students to return to the classification of numbers from the starter activity (or choose another suitable example) and create a Venn diagram showing where the numbers go when three of the classification groups are chosen for the circles in the diagram.

Alternatively, discuss how Venn diagrams could be modified to take into account four (or more) different characteristics. (Can this be done in only two dimensions? A three-dimensional Venn diagram with four intersecting spheres will certainly make the students think but how about a four-dimensional Venn diagram with five intersecting *hyper*spheres?)

Lessons 4 and 5 – Introduction to vectors

Textbook pages 289–295

Objectives

E7.1: Describe a translation by using a vector represented by e.g. $\begin{pmatrix} x \\ y \end{pmatrix}$, $\overrightarrow{AB}$ or **a**. Add and subtract vectors. Multiply a vector by a scalar.

E7.3: Represent vectors by directed line segments. Use the sum and difference of two vectors to express given vectors in terms of two coplanar vectors.

Starter

Give the students a coordinate pair, say (2, 1) and ask them to describe how they would get to another coordinate pair, say (4, 6). Repeat this several times (include negative coordinates).

Lesson commentary

- By linking in to the starter activity (and work done previously on translations), describe to students the concept of a vector (having a size/magnitude *and* a direction) and explain that vectors can be represented as directed arrows which point in the given direction. Discuss the process of joining a number of different vectors together from their end points to represent vector addition. Further questions that can be asked at this stage include 'What happens when we double vector **a**?' (we get 2**a**) and 'How might we represent **a** – **b**?'.

- Explain to students that when we are given a geometrical problem involving vectors, it is just like giving directions to someone on how to get from their starting point to a fixed destination, but that we must always travel along *known* vectors. Provide some simple examples for students to attempt before gradually increasing the level of complexity. Students, possibly working in groups or pairs, if appropriate, could be provided with different questions in order to test their relative abilities.

- There are two long exercises associated with this work and most students will need time to consolidate this work fully. The second lesson can therefore be used to provide additional examples, support for weaker students and an opportunity for more able students to attempt more complex questions.

Plenary

Provide students with a grid similar to that at the start of exercise 5 (or use this grid itself) and give them a quiz on how to get from 'A' to 'B'.

Exercise commentary

Questions 1 to 26 of exercise 5 are all basic practice in navigating around a grid while **questions 27 to 38** work the same concept backwards. These can be used as much as required but could be skipped through quickly by class question and answer. The remaining problems are a precursor to the more formal vector geometry covered later on.

The questions in exercise 6 introduce the idea of midpoints and again lead on to the more formal vector geometry. These could be used sparingly at this stage or given as extension questions for more able students.

Lessons 6 and 7 – *Column vectors*

Textbook pages 295–299

Objectives

E7.1: Describe a translation by using a vector represented by e.g. $\begin{pmatrix} x \\ y \end{pmatrix}$, $\overrightarrow{AB}$ or **a**. Add and subtract vectors. Multiply a vector by a scalar.

E7.3: Represent vectors by directed line segments. Use the sum and difference of two vectors to express given vectors in terms of two coplanar vectors. Use position vectors.

Starter

Identify the coordinates: Provide students with a number of clues as to the location of the 'treasure' and ask them to work out where it is. Clues could include 'the x coordinate is a multiple of 3', 'the y coordinate is less than 10' and 'the sum of the coordinates is prime'.

Lesson commentary

- Students should be familiar with the idea of column vectors since they will have used them in their work on translation. A quick revision activity might be all that is needed before linking this back to the previous lesson. Discuss generalised methods for adding, subtracting and multiplying vectors when they are given numerically. Discuss the concept of parallel vectors (one being a multiple of the other – same direction).

- Students could be provided with a number of examples at this stage although many will be happy to attempt some questions straight away. Encourage students to work together and to form discussions if appropriate.

- The second lesson will provide students with sufficient time to fully consolidate this work on column vectors.

Exercise commentary

Exercise 7 contains plenty of basic practice and can be used more or less extensively depending on the ability of the students. **Questions 37 and 38** deal with the concept of parallel vectors while **questions 39 and 40** ask students to draw resultant vectors.

In exercise 8, students are asked to tackle a series of geometrical problems on coordinate grids. These are certainly more conceptually difficult and the examples from **question 7 onwards** can be used as extension work.

Plenary

Provide students with two further column vectors, for example

$\mathbf{a} = \begin{pmatrix} 4 \\ 7 \end{pmatrix}$ and $\mathbf{b} = \begin{pmatrix} -2 \\ 3 \end{pmatrix}$, and ask them to work out vector sums such as

$\mathbf{a} + \mathbf{b}$ and $3\mathbf{b}$.

Lesson 8 – *Modulus of a vector*

Textbook pages 299–301

Objectives

E7.1: Describe a translation by using a vector represented by e.g. $\begin{pmatrix} x \\ y \end{pmatrix}$, $\overrightarrow{AB}$ or **a**. Add and subtract vectors. Multiply a vector by a scalar.

E7.3: Calculate the magnitude of a vector $\begin{pmatrix} x \\ y \end{pmatrix}$ as $(\sqrt{x^2 + y^2})$.

Starter

Revision of Pythagoras' theorem: Provide students with examples of right-angled triangles where they are required to work out the length of the hypotenuse when given the shorter two side lengths.

Lesson commentary

- Explain to students that the two numbers in a vector can be referred to as the *components* of the vector and just describe the number of units across and up/down. Show how the vectors could be represented as right-angled triangles with the vector as the hypotenuse.

- Ask students how they would work out the length of the vector (or modulus/magnitude). Expect most students to quickly identify the need for Pythagoras' theorem and provide them with examples which use this as appropriate.

- Students can then complete consolidation questions on this topic using examples from the textbook.

Exercise commentary

Finding the modulus of a vector is tested in exercise 9. The first six questions are basic practice before students are asked to link together the arithmetic of vectors with finding the modulus. **Question 13 onwards** might be set aside for stretching more able students.

Plenary

Provide students with several more vectors and ask them to work out the modulus (using a calculator, if necessary) and write down the answers in their exercise books.

Lessons 9 and 10 – *Vector geometry*

Textbook pages 301–304

Objectives

E7.1: Add and subtract vectors. Multiply a vector by a scalar.

E7.3: Represent vectors by directed line segments. Use the sum and difference of two vectors to express given vectors in terms of two coplanar vectors. Use position vectors.

Starter

Give the students two column vectors and ask them to quickly write down the answers to certain sums such as $\mathbf{a} + \mathbf{b}$, $2\mathbf{a} - 3\mathbf{b}$, etc. They can write their answers on mini-whiteboards, if available, or write them in their exercise books.

Lesson commentary

- Most students find the ideas of vector geometry hard to grasp but the previous lessons should have given them a better conceptual understanding of vectors as 'signposts' pointing the way. Repeat the concept that when travelling around the various diagrams, it is essential that you only travel along sides where the vectors are *known* (this could include those calculated in previous parts of the question).

- Ask the students to work in pairs or small groups and provide them with appropriately differentiated examples. Ask them to discuss how they might solve the problems using the ideas of the previous two lessons. This whole-class approach should help them to develop an understanding of the issues involved. Students can then be provided with further examples to attempt either on their own or still within their groups.

- The second lesson can be used to ensure that students complete a suitable range of examples from the exercise in the textbook. Further support and worked solutions can be provided as appropriate.

Exercise commentary

The questions in exercise 10 get increasingly more complicated and therefore students should be directed to complete questions that are appropriate to their ability. Emphasise the need for *proof* when making geometrical deductions such as coplanarity and parallel lines.

Plenary

Provide a further example and ask students to write down the answers to various questions asked about the diagram. Include midpoints but avoid ratios.

Lesson 11 – *Introduction to functions*

Textbook pages 305–307

Objectives

E2.9: Use function notation, e.g. $f(x) = 3x - 5$, $f : x \rightarrow 3x - 5$, to describe simple functions.

Starter

Think of a number: Give the students a range of 'I think of a number ...' type problems and ask them to work out the number that you originally started with.

Lesson commentary

- By referring back to both the starter activity and the work done previously on equations and expressions, discuss the concept of a function. Function machines can be revised if necessary. Introduce the notation for functions, both $f(x)$ and $f: \longmapsto x$.

- Demonstrate that when evaluating functions such as $f(2)$, all we are really doing is substituting into the expression for the given function. Students can then spend time getting used to using the notation and the ideas behind functions by working through a series of consolidation questions. Encourage students to work together and active discussion throughout, if appropriate.

Exercise commentary

Questions 1 and **16 to 20** of exercise 11 are evaluative where students must substitute numbers into given functions. **Questions 2 to 15** require students to draw flow diagrams to represent given functions whereas **questions 21, 23 and 25** lead onto the idea of an inverse function. **Questions 26 to 28** are more conceptually difficult and could be set aside for extension work.

Plenary

Give students another example of a function and ask them to determine the value(s) of x which give a certain value to the function. For example, if the function $f(x) = 2x + 3$ is equal to 7, what is the value of x?

Lessons 12 and 13 – *Inverse and composite functions*

Textbook pages 307–310

Objectives
E2.9: Use function notation, e.g. $f(x) = 3x - 5$, $f : x \rightarrow 3x - 5$, to describe simple functions. Find inverse functions $f^{-1}(x)$. Form composite functions as defined by $gf(x) = g(f(x))$.

Starter
Provide students with a range of simple formulae of the form $y = f(x)$ and ask them to make x the subject of the formula.

Lesson commentary
- It seems sensible to split this topic into two parts and deal with inverse and composite functions separately but since the textbook exercise covers both aspects, the lesson commentary includes both. The second lesson can be used to split the topic or to consolidate the work on inverse and composite functions taught together in the first lesson.

- Explain that what students have been doing as part of the starter activity is to find the inverse function, and introduce the formal notation used to describe the inverse function. Students may initially be confused by the use of x on both sides of the equality sign when working with functions, but an approach that might help is to turn the function into the form $y = ...$, rearrange to make x the subject, and then replace the y with an x in the completed inverse function expression. Students should be given the opportunity to work through several examples of this and encourage active discussion throughout.

- By providing two different functions $f(x)$ and $g(x)$, ask students to substitute a value into f before substituting the answer into g. Explain that what they have done is to work out the value of the composite function gf. Explain that the order of the letters is important since the letter closest to the x is the first function to be applied.

- Demonstrate that composite functions can be determined algebraically by substituting the expression for the first function in place of the x in the second. Simplification is usually necessary to get the function into a neat form but this is less important than understanding the processes involved at this stage.

- Allow students time to consolidate their work.

Exercise commentary
The first part of exercise 12 deals with composite functions and the second part with inverse functions. These could be done in reverse order if the lesson commentary is followed. **Questions 2, 5 and 8** are more conceptually difficult and could be set aside for extension work. **Question 22** is an investigative question looking at calculator buttons and could be left out.

Plenary
Provide students with two further functions f(x) and g(x) and ask them to substitute numbers into them, find the inverses and find composite functions such as fg and gf. They can write down the answers in their exercise books for checking at the end.

Lesson 14 – Transformations

Textbook pages 310–320

Objectives

E7.2: Reflect simple plane figures. Rotate simple plane figures through multiples of 90°. Construct given translations and enlargements of simple plane figures. Recognise and describe reflections, rotations, translations and enlargements.

Starter

Ask students to give examples of capital letters which have one/two/no lines of symmetry and those which have rotational symmetry.

Lesson commentary

- Students should be very familiar with simple transformations at this level and this lesson could be used for revision of the key ideas. Ask students to write down, in their own words, the basic principles of applying known transformations such as rotation, reflection, translation and enlargement. Encourage a complete description and the minimum requirements to fully describe these transformations. Ask students to work with a partner on these definitions, if appropriate, and check that they both fully understand the processes. Get selected students to report back to the class to determine the accuracy of their definitions/instructions.

- Most students will then benefit from a mixed exercise or series of examples which test their understanding of these transformations, rather than exercises which look at each one in turn. If necessary, allow further lessons to be used to ensure a full understanding.

Exercise commentary

Exercises 13 to 18 look at each of the transformations in turn and could be reserved for further consolidation if students are struggling with any of these ideas. More able students should be able to start at exercise 19 which involves all of the basic transformations in a mixed exercise.

Plenary

Students could challenge a partner to perform and discuss given transformations on a shape of their choosing. Ensure the transformations are sensible though (*not* something like 'reflect in the line $y = 3x + 2$'). Alternatively, a short revision test can be used to check understanding.

Lessons 15 and 16 – *Combined transformations*

Textbook pages 320–324

Objectives
E7.2: Reflect simple plane figures. Rotate simple plane figures through multiples of 90°. Construct given translations and enlargements of simple plane figures. Recognise and describe reflections, rotations, translations and enlargements.

Starter

Ask students to think about the point (2, 1) and then describe a number of transformations that students should imagine performing on the point. These could include reflections and rotations as well as translations. The aim is for students to write down the coordinates of the final image (intermediate steps can be recorded but discourage use of a coordinate grid).

Lesson commentary

● Revise the idea of right to left notation as encountered in the work on matrices and discuss the idea that a series of transformations can be described in the same way using suitable letters to denote the transformations.

● Provide students with a number of transformations and ask them to draw a simple shape in the first quadrant of a set of coordinate axes. They can then proceed to apply the transformations in a given order to their shape. Students could work in pairs for this activity and challenge each other to do different transformations or the activity can be done as a class with students working individually.

● Discuss repeated transformations and inverse transformations, emphasising the need for clear and accurate use of the notation at all times. Students could be challenged to write down the inverses to a number of familiar transformations such as vector translations and simple reflections and rotations.

● The second lesson can be used for additional consolidation of this work and the completion of questions from the exercises. More able students will benefit from extension work while weaker students can get further support.

Exercise commentary

Exercise 20 is basic practice in applying combinations of transformations to a given point. Students are advised to use a diagram but more able students may be able to visualise the point moving in space (see the starter activity).

In exercise 21, students are transforming shapes rather than points and the use of a grid here will be very useful: it is much more difficult to think about three points moving relative to each other under a combination of transformations. **Question 6** focuses on finding inverses before **questions 7 to 9** use these inverse transformations on given objects.

Plenary

Students could be instructed to match a series of objects and images with their transformations. These could be provided on a worksheet or on a screen using geometry software or graph-drawing software.

The revision and examination-style exercises can be used for further practice as appropriate.

CHAPTER 9
STATISTICS

Lessons 1 and 2 – *Data display*

Textbook pages 332–343

The support worksheet for Chapter 9 may be useful to support lesson 4.

The challenge worksheet for Chapter 9 may be useful to support lesson 4.

Objectives

E9.1 Collect, classify and tabulate statistical data.

E9.2: Read, interpret and draw simple inferences from tables and statistical diagrams.

E9.3: Construct and interpret bar charts, pie charts, stem-and-leaf diagrams, simple frequency distributions, and histograms with equal and unequal intervals.

Starter

Provide students with a number of division sums which lead to decimal answers. Examples could include sums such as $9 \div 4$ and $14 \div 5$. Responses could be on mini-whiteboards or written into exercise books for checking at the end.

Lesson commentary

- Students should be familiar with basic bar charts, pie charts and frequency polygons and therefore the focus for this lesson should be on the accurate construction of histograms.

- Explain to students that histograms are similar to bar charts but should be used for continuous data such as height and weight. Explain the key points: there should be no spaces between the bars, and they should start and finish at the lower and upper class boundaries for the group of data. It is commonly thought that the height of the bar should represent the frequency whereas, in fact, (for unequal class intervals) it is the area of the bar that represents the frequency. Introduce the idea of *frequency density* and demonstrate an example for the class. Encourage the students to add a column to their tables for the frequency density calculation before drawing the final diagram.

- Students could collect their own data (heights, for example) before drawing their own histograms. Encourage them to think about the widths of the classes carefully to get a thorough representation.

Exercise commentary

Exercise 1 contains basic practice in drawing and interpreting bar charts and pie charts and could be omitted if the group are happy with these types of statistical diagram.

Exercises 2 and 3 cover stem-and-leaf diagrams and frequency polygons respectively, and again these could be omitted.

Exercise 4 has a range of questions in which students are expected to draw histograms and these can be used for consolidation. In **questions 6 and 7**, the class boundaries need to be considered carefully before proceeding to draw the histograms.

Plenary

Provide students with a further example and ask them to complete the diagram, possibly working in pairs if appropriate to check each other's working.

Lessons 3 and 4 – *Mean, median and mode*

Textbook pages 344–349

Objectives

E9.4: Calculate the mean, median, mode and range for individual and discrete data and distinguish between the purposes for which they are used.

E9.5: Calculate an estimate of the mean for grouped and continuous data. Identify the modal class from a grouped frequency distribution.

Starter

Ask students to write down the mean, median and mode for simple sets of discrete data (odd numbers of values *and* even numbers of values).

Lesson commentary

- Students should be happy to work out the mean, median and mode for discrete data sets (see starter activity). The emphasis here should be on grouped frequency tables.

- Data could be collected from the group that is both discrete (but useful to group) such as number of siblings for example, and continuous such as height or hand span. This data can then be used to demonstrate the methods for finding the mean, median and mode from grouped data.

- Begin with the basic approach for grouped discrete data, emphasising that, for the mean, the *total number* should be divided by the *total frequency*, not the number of groups. The median should be worked out by counting up half of the total frequency (+1) and the mode is the value of the most common class.

- Extend the work on mean to grouped continuous data where the use of the mid-interval value is required to calculate an *estimate* of the mean. Discuss the concept of a modal *class* and point out that the median cannot be accurately determined in this case (more able students might be able to work on the idea of linear interpolation but this is not required).

- Students can then practise further examples of both types.

- The second lesson will allow sufficient time for consolidation of this topic and could also be used to provide support to less able students on simple measures of average.

Exercise commentary

Exercise 5 begins with routine practice in working out the mean, median and mode from simple data sets before moving on to discrete grouped data. From **question 16 onwards**, the problems are more complicated and these could be used for extension work.

Exercise 6 requires students to work out the mean and/or the modal class for the grouped data sets given. These questions are routine practice of this skill.

Plenary

Students could be asked to come up with sets of data which have certain characteristics. For example, four numbers with a mean of 5, a median of 6 and a mode of 4.

Lessons 5 and 6 – *Scatter graphs*

Textbook pages 349–354

Objectives

E9.1: Collect, classify and tabulate statistical data.

E9.3: Construct and interpret scatter diagrams.

E9.7: Understand what is meant by positive, negative and zero correlation with reference to a scatter diagram.

E9.8: Draw, interpret and use lines of best fit by eye.

Starter

Ask students to suggest whether they think certain pairs of data are connected. For example, height and weight, number of ice creams sold and the temperature, test scores in Mathematics and English or shoe size and hand span. Ask them to briefly justify their responses.

Lesson commentary

- Follow on from the starter by formalising some of the language of bivariate data analysis. Ask students to discuss what they think is meant by (and the differences between) correlation and causality.

- Introduce the idea of a scatter graph to show the relationship between two sets of data and discuss what the spread of points might look like for each of the examples already discussed.

- Students can collect their own data here (such as the measurements suggested in question 1 of exercise 1) and draw a scatter graph for this data.

- Formally introduce the ideas of correlation and lines of best fit. Ask the students to describe the type (and strength) of correlation they would expect for the examples already discussed and their own graphs. They could put a line of best fit on their own scatter graphs at this point.

- Further practice in drawing and interpreting scatter graphs can then be given as appropriate. The second lesson will enable students to have sufficient time to effectively consolidate this work.

Exercise commentary

Question 1 in exercise 7 could be used to introduce the topic to the class (as described in the lesson commentary). **Question 2** asks students to draw further scatter graphs and interpret the correlation. **Questions 3 to 7** ask them to draw a line of best fit onto their diagrams and use this to estimate various values.

Plenary

Ask students to describe three situations (other than the ones discussed as part of the lesson), one with a strong positive correlation, one with a strong negative correlation and one with no correlation. Ask for volunteers to share some of their ideas with the class.

Lesson 7 – *Box-and-whisker plots*

Textbook pages 354–355, 360–361

Objectives
E9.2: Read, interpret and draw simple inferences from tables and statistical diagrams. Compare sets of data using tables, graphs and statistical measures. Appreciate restrictions on drawing conclusions from given data.

E9.6: Construct and interpret box-and-whisker plots.

Starter
Revise the ideas of the median, upper and lower quartiles and interquartile range. Demonstrate how to use these to construct a box-and-whisker plot. Tell the class that these diagrams are sometimes just called 'box plots'.

Lesson commentary
- Once students are able to construct box-and-whisker plots, tell them that the main purpose of these diagrams is to be able to compare more than one set of data. Use questions 2 and 4 from exercise 10 (pages 360 and 361) to demonstrate this.

- Students need to know that when comparing two box-and-whisker plots, they must compare the medians, then compare the spread of the data.

Exercise commentary
Exercise 8 is very short because there is very little to understand about the basics of constructing a box-and-whisker plot.

Exercise 10 is about how we can compare data sets, and there are two questions (**questions 2 and 4**) in that exercise that involve box-and-whisker plots. **Question 3** could be used to show how a box-and-whisker plot can be constructed from a cumulative frequency graph, so should be saved until students have learned about cumulative frequency graphs in the next lessons.

Plenary
Look at question 1 on page 360, which is about pie charts. Ask students, 'Why are there so many different kinds of graphs?', 'What are the purposes of each one?', 'Why would a box-and-whisker plot not be suitable for question 1?'.

Lessons 8 and 9 – *Cumulative frequency*

Textbook pages 356–359

Objectives
E9.6: Construct and use cumulative frequency diagrams. Estimate and interpret the median, percentiles, quartiles and interquartile range.

Starter
Provide students with a starting number, say 5, and ask them to successively add other numbers to it. Repeat several times, always asking for the final number to be recorded.

Lesson commentary

- A data set such as that in the example in the textbook or one which has been collected by the class can be used to introduce the idea of a cumulative frequency diagram. Start from the point of a grouped frequency table and demonstrate how students can add successive frequencies to generate a third column (cumulative frequency). Explain that the cumulative frequency diagram can then be plotted using the *upper class boundary* as the x coordinate and the cumulative frequency as the y coordinate. Indicate the expected shape of the final curve (an *ogive*).

- Students will benefit from practising further examples where they have to draw the diagrams before moving on to using cumulative frequency diagrams to work out estimates for the median and quartiles.

- When the median and quartiles are considered, emphasise that they are just estimates but can be fairly accurate for a large data set. Introduce the notion of *interquartile range* as the difference between the upper and lower quartiles. Students should see this as a measure of spread (similar to the range but not affected by outlying values).

- The second lesson will enable students to consolidate this work fully and allow time to be spent on the accurate drawing of diagrams.

Exercise commentary
Students use the graphs in **questions 1 and 2** of exercise 9 to work out estimates for the median and quartiles while in **questions 3 to 7** they must first draw the diagrams before working out estimates for the median and quartiles. In **question 8**, the mean is reintroduced while **questions 9 and 10** look at comparing data using the median and interquartile range.

Plenary
Students could be presented with one further example and asked to estimate the median and quartiles. A more formal treatment of deciles and percentiles could also be discussed at this point.

The revision and examination-style exercises can be used for further practice as appropriate.

CHAPTER 10
PROBABILITY

Lessons 1 and 2 – Simple probability

Textbook pages 366–373

Objectives

E8.1: Calculate the probability of a single event, as either a fraction, decimal or percentage.

E8.2: Understand and use the probability scale from 0 to 1.

E8.3: Understand that the probability of an event occurring = 1 – the probability of the event not occurring.

E8.5: Calculate the probability of simple combined events.

Starter

Complements to one: Give the students a series of numbers between 0 and 1 which have two decimal places (0.34 for example). Ask the students to write down the complement of this number (the number to which it would be added to make a total of 1).

Lesson commentary

- Students are likely to be very familiar with the language of probability so much of the initial discussion could be done quickly as a class. Ask informally (likely, unlikely, impossible, etc.) the probability of some events happening: the likelihood of it raining tomorrow, the likelihood of someone in your class having a birthday in the next month.

- Discuss how these probabilities can be converted into a number and discuss the probability scale. Formalise the idea of a probability being 'the number of successful outcomes divided by the total number of outcomes' (expressed as a fraction). Provide a number of examples (balls in bags, rolling dice, etc.) and ask the students to give the probability of several events happening. Include impossible and certain events.

- Formalise the idea of complementary probabilities (the probability of something *not* happening).

The support worksheet for Chapter 10 may be useful to support lessons 1 and 2.

The challenge worksheet for Chapter 10 may be useful to support lesson 5.

Exercise commentary

The questions in exercise 1 test the students' understanding of basic probability and uses listing of outcomes to guide them in several cases. Most students should be able to answer these questions easily and they could be used during whole-class discussion rather than set as a formal exercise.

Exercise 3 looks at exclusive and independent events.

- Discuss the idea of *expectation*. If you know the theoretical probability of something happening, can you work out the expected number of successes in a given number of trials?

- Describe the principle of *mutually exclusive* events and refer to the starter activity for a series of examples.

- Discuss the link between mutually exclusive outcomes and the idea of complementary probabilities; provide students with further examples as necessary.

- Discuss independent events and ask students to come up with examples of events which are independent and dependent (head on coin/six on die against miss bus/late to school).

- The second lesson can be used to consolidate this work fully through completion of exercise questions from the textbook.

Plenary

Provide the students with further examples and ask them to calculate probabilities based on mutually exclusive and independent events.

Lesson 3 – *Relative frequency*

Textbook pages 370–371

Objectives
E8.4: Understand relative frequency as an estimate of probability.
Expected frequency of outcomes.
E9.1: Collect, classify and tabulate statistical data.

Starter

Ask students to draw two 6 by 6 grids, with numbers from 1 to 6 across the top and down the side. Inside the first grid, in each box, they should write the sum of the number at the top of the column and the number at the start of the row. In the second grid, they should write the products of the numbers. These grids could be pre-prepared for weaker students.

Lesson commentary

- Working in pairs, ask students to roll two dice multiple times, each time noting down the sum and the product of the two dice. After a few minutes, collect everyone's data together. Establish the total number of dice rolls for the group, the total number of times that the sum was an even number and the total number of times that the product was an even number. The proportion of even sums and products should roughly match the proportions in the tables.

- Explain that if we did not have the tables, we could estimate the probabilities using the results of the experiments.

- Explain that relative frequency is the way in which many real-life probabilities are calculated, as most things cannot be analysed in simple mathematical ways.

- Discuss the fact that the more times you repeat the experiment, the more accurate the probabilities will be.

Exercise commentary

Exercise 2 contains two experiments related to relative frequency.
Question 1 involves rolling a biased dice. **Question 2** involves drawing a conclusion based on analysing the data in the table.

Plenary

Buffon's Needle is a famous problem in geometry, whose solution allows you to estimate the value of pi using relative frequency. The students could actually try this out for themselves, or it could just be the subject of a class discussion.

Lessons 4 and 5 – *Tree diagrams*

Textbook pages 374–378

Objectives
E8.5: Calculate the probability of simple combined events, using possibility diagrams and tree diagrams.

Starter

Give students the following scenario: 'There are 7 balls in a bag, 4 green and 3 red.' Ask students to write down the probability of picking, say, a red ball $\left(\frac{3}{7}\right)$. Ask them to then write down the probability of picking a red ball the *next* time, assuming the first one is not returned to the bag $\left(\frac{2}{6} \text{ or } \frac{1}{3}\right)$. Continue this line of questioning or change the example.

Lesson commentary

- Use the example in the starter activity (or another suitable example) to describe to students the process of drawing a tree diagram to show successive events, such as taking out balls from a bag or rolling dice. Include examples which have replacement as well as those which do not. Most students find the construction of the tree diagrams a relatively straightforward exercise while some struggle to work out probabilities, especially in cases without replacement. Encourage students to work together if appropriate, and ensure students check their probabilities before using them to perform any calculations.

- Examples of how the diagram can be used to calculate the probabilities of combined events can then be given to the students before they go on to try some examples of their own.

- The second lesson can be used to complete further questions from the exercise in the textbook to ensure that students have effectively consolidated this topic.

Exercise commentary

The first two questions in exercise 4 involve replacement while the next three do not. **Questions 4 and 5** both have three possible outcomes at each stage so ensure students are comfortable with drawing tree diagrams with more than two branches at each level. **Questions 6 to 8** concern three events. The remaining questions in the exercise are a mixture of problem-solving questions and can be used as appropriate.

Plenary
Students could be provided with a further example (projected onto a screen, for example) and asked questions about it. Responses could form part of a small test or be written into exercise books for checking at the end.

Lesson 6 – Probability from Venn diagrams

Textbook pages 378–381

Objectives
E8.5: Calculate the probability of simple combined events using Venn diagrams.

Starter
Ask the class to tell you everything they can remember about Venn diagrams. Look back at pages 280–285 if necessary. Use example 3 on page 378 to demonstrate how probabilities can be calculated using Venn diagrams.

Lesson commentary

- The first two questions in exercise 5 should be fairly straightforward and the students should not need much guidance.

- The aim of questions 3 and 4 is that students should be able to construct their own diagrams. Encourage them to do this by themselves if possible.

- Although question 6 contains algebra, it is very simple algebra and students should not be put off by this.

Exercise commentary

In **questions 1 and 2** of exercise 5, there are only two overlapping regions inside the diagram, and the diagrams are provided. In **questions 3 and 4**, the diagrams are not provided. **Questions 5 and 6** contain three overlapping regions, with **question 6** being more difficult because it involves algebra. **Question 7** contains the more difficult set notation. **Questions 6 and 7** in this exercise should be considered extension work for more able students.

Plenary
Is there an easier way of writing $P(A' \cap B')$?
Ask students to draw a diagram and shade this region. They should notice it is the same as $P(A \cup B)$. Teach the class about DeMorgan's laws and how they can help simplify things like this.

Lesson 7 – *Conditional probability*

Textbook pages 381–383

Objectives
E8.6: Calculate conditional probability using Venn diagrams, tree diagrams and tables.

Starter
Discuss the idea of independent events. Can students think of two events that are **not** independent? In other words, can they think of something whose probability changes if something else has happened? For example, are they more likely to be late for school if it is raining?

Lesson commentary
- Present the formula for P(A|B) early on in the lesson and familiarise the students with this notation.

- It is important that students know how to calculate a conditional probability from Venn diagrams, tree diagrams and two-way tables. The differences and similarities between these methods should be discussed with the class.

- It is important that students understand that a conditional probability is essentially a probability where the denominator is changed as a result of something that happened previously.

- The formula for P(A|B) can be easily understood when visualised using one of the diagrams. The formula should not be seen as something that replaces the diagrams, but rather as something to be used only under certain circumstances (e.g. when all the separate parts that go into the formula have been calculated previously).

Exercise commentary

Students should be able to complete **questions 1 and 2** of exercise 6 without drawing a diagram. **Question 3** uses a two-way table. **Question 4** uses a tree diagram. **Question 5** requires students to draw their own tree diagrams. Discuss with the class why a tree diagram is the most appropriate diagram for this question. **Questions 6 and 7** use Venn diagrams.

Plenary
Discuss the following question with students: 'How can we use P(A|B) to check whether events A and B are independent?' A and B are independent if P(A) = P(A|B). Ask: 'What is the logic behind this argument?'

The revision and examination-style exercises can be used for further practice as appropriate.

CHAPTER 11
INVESTIGATIONS, PRACTICAL PROBLEMS, PUZZLES

This chapter contains a number of interesting investigations and other problems. These tasks could be spread throughout the course as stand-alone lessons (or series of lessons) or they could be tackled at the end of the course. They could also be used as extension and enrichment activities. The key mathematical elements to each one are listed below.

1 *Opposite corners* This investigation requires students to be able to multiply one- and two-digit numbers, and then generalise numbers in a square grid using algebra. The ability to carry out algebraic manipulation is essential for high marks.

2 & 3 *Scales and Buying stamps* These are both logic-based activities which require a good degree of thought through reasoning.

4 *Frogs* This is a puzzle testing a student's ability to be logical and recognise patterns.

5 *Triples* A number investigation requiring a logical approach and ability to generalise.

6 *Mystic rose* An investigation which leads to triangle number sequences.

7 *Knockout competition* Logical deduction leading to a surprisingly simple result.

8 *Discs* Similar to *Frogs*, this tests logical skills and recognising patterns.

9 *Chessboard* Spatial awareness and visualisation is required here.

10 *Area and perimeter* Skills with area and perimeter are essential but a good degree of logic is required too. Good spatial awareness is needed.

11 *Happy numbers (and more)* An investigation which uses simple arithmetic skills.

12 *Prime numbers* Knowledge of prime numbers is essential here, as is an element of logic.

13 *Squares* Pythagoras' theorem and a knowledge of surds will be useful here.

14 *Painting cubes* Good visualisation skills and the ability to generalise algebraically are important, as is an understanding of sequences and nth term formulae.

15 *Final score* This investigation is logic-based and requires students to use ideas such as systematic listing and algebraic generalisation.

16 *Cutting paper* A knowledge of mathematical similarity is required here.

17 *Matchstick shapes* Knowledge of nth term formulae and sequences are required.

18 *Maximum box* Multiplication of two-digit numbers is necessary (although calculators are permitted). For a higher level generalisation, good algebraic manipulation skills are essential.

19 *Digit sum* Recognising patterns and logical deduction are essential.

20 *An expanding diagram* This investigation is based on the work students do on sequences and generalising for the nth term formula.

21 *Fibonacci sequence* Investigating one of the most famous sequences of mathematics.

22 *Alphabetical order* An introduction into permutations and combinations. Again logical sequencing is an important part of it.

23 *Tiles* An understanding of sequences and skills in generalisation are necessary.

24 *Diagonals* Students need an appreciation of shape and space, good sequencing skills and a logical mind. They are looking for patterns and therefore the ability to generalise in algebraic terms is useful.

25 *Biggest number* A calculator exercise which uses logical reasoning.

26 *What shape tin?* Knowledge of cylinder volume and surface area are required.

27 *Spotted shapes* An investigation into Pick's theorem. Good spatial reasoning and logical deduction are needed here.

28 *Stopping distances* This looks at formulae connecting speed and distance.

29 *Maximum cylinder* Knowledge of cylinder volume and surface area are required.

These practical problems are good for encouraging students to think about mathematics in context.

1 *Timetabling* This is a logic-based activity which requires good spatial awareness and the ability to model using restrictions. An appreciation of trial and improvement methods is essential.

2 *Hiring a car* Students will need to work with money, use ideas of proportion, and be able to carry out multi-stage arithmetical calculations.

3 *Running a business* A financial management problem verging on linear programming. A logical approach is vital.

4 *How many of each?* A linear programming problem set in context. Students need to have completed the work on inequalities and linear programming (or use logical investigation).

The puzzles and experiments in section 11.3 are extra to what is required by the course. They make interesting homework activities, extension activities, or end-of-term activities.

1 *Cross numbers* These are purely logic-based. No mathematical skill is required.

2 *Estimating game* A game testing both multiplication skills and the ability to estimate the answers to calculations.

3 *The chessboard problem* A logic problem.

4 *Creating numbers* Good basic number skills and the correct application of the rules of BIDMAS are essential here.

5 *Pentominoes* A shape and space investigation with space-filling problems later on.

6 *Calculator words* These are often problematic with modern calculator displays, but can be used with a little 'imagination' to test calculator skills.

The multiple-choice revision tests (Chapter 12) can be used for consolidation and revision at the end of the course (or at other appropriate points during it).